T5-DGY-727

The Mechanic's Guide to
ELECTRONIC EMISSION CONTROL AND TUNE-UP

Larry W. Carley

Bob Freudenberger

Prentice-Hall, Inc., Englewood Cliffs, New Jersey 07632

Library of Congress Cataloging-in-Publication Data

Carley, Larry W.
  The mechanic's guide to electronic emissions control and tune-up.

  Includes index.
   1. Automobiles—Pollution control devices. 2. Automobiles—Pollution control devices—Maintenance and repair. I. Freudenberger, Bob. II. Title.
TL214.P6C37 1985   629.2′528′0288   85-16989
ISBN 0-13-569815-4

Editorial/production supervision: Fred Dahl
Cover design: Joseph Curcio
Manufacturing buyer: Gordon Osbourne

Printed in the United States of America

10 9 8 7 6 5 4 3 2 1

ISBN   0-13-569815-4   01

Prentice-Hall International (UK) Limited, *London*
Prentice-Hall of Australia Pty. Limited, *Sydney*
Prentice-Hall Canada Inc., *Toronto*
Prentice-Hall Hispanoamericana, S.A., *Mexico*
Prentice-Hall of India Private Limited, *New Delhi*
Prentice-Hall of Japan, Inc., *Tokyo*
Prentice-Hall of Southeast Asia Pte. Ltd., *Singapore*
Editora Prentice-Hall do Brasil, Ltda., *Rio de Janeiro*
Whitehall Books Limited, *Wellington, New Zealand*

to the
hard working mechanics
of America

Contents

Preface, vii

1
Overview, 1

2
Basic Pollutants: HC, CO, and NO_x, 11

3
Positive Crankcase Ventilation:
Airing Out the Engine, 20

4
Evaporative Emission Controls, 29

5
Heated Air Intake:
Blending to the Engine's Taste, 40

6
Air Injection: Force Feeding O_2, 48

7
Catalytic Converters, 60

8
Exhaust Gas Recirculation (EGR), 70

9
How Carburetion Affects Emissions, 88

10
How Ignition Affects Emissions, 101

11
**Other Factors
That Influence Emissions, 121**

12
Emissions Testing, 129

13
Troubleshooting Emissions Problems, 147

14
**Troubleshooting Drivability
Problems That Are Emissions Related, 163**

15
Tuning-Up Emission-Controlled Engines, 171

16
Emission System Modifications, 186

Glossary, 197

Appendix A: Automotive Acronyms, 208

Index, 213

Preface

This book is not an emissions control service manual. If you want service specs, vacuum routing diagrams, or step-by-step instructions for replacing emissions components, you should use one of the many repair manuals that are available.

The purpose of this book is to give you what most repair manuals leave out—the basics of how the various emissions control systems work and how to troubleshoot them.

The subject of emissions control is very complex. So it's essential for today's mechanic to understand the theory behind the systems on which he's working. Only then can he accurately diagnose and repair emissions-related problems. The old trial-and-error method of repair just doesn't cut it anymore. You have to understand what you are doing and why.

We've written this book from a generic point of view so that the basic theory can be applied to virtually every vehicle. In some instances, we've used a specific example to explain a certain point. In others, we've listed the basic variations on a common theme that are currently being used. In any event, you should be able to use this book as a reference in conjunction with your repair manual when diagnosing and servicing emissions problems.

ACKNOWLEDGMENTS

The authors wish to thank the vehicle manufacturers, equipment suppliers and the National Center for Motor Vehicle Emissions Control Training for technical information and illustrations used in this book. We would also like to thank the United States Congress, the Environmental Protection Agency, and scores of unnamed automotive engineers who are responsible for creating today's complex emission control systems, and without whose efforts a book such as this would never have been needed.

1

Overview

This is a book about emission control. Its purpose is to help you understand how emission control systems work so that you can accurately diagnose and repair emission-related problems. Each chapter deals with a different aspect of emission control. By the time you finish reading this book, you should know the theory behind today's emission control systems. This will enable you to troubleshoot problems with a greater degree of confidence. Of course, you will still need a good shop manual for application specifics, performance specifications, vacuum hose routing diagrams, replacement procedures, and so on. But with an understanding of the how's and why's of emission control and the effects it has on fuel economy, performance, starting, and drivability, repairing emission-related problems should be no more troublesome than any other aspect of automotive service work.

WHY THINGS ARE THE WAY THEY ARE TODAY

Since the early 1970s, automotive emission control has undergone an evolution that would make Darwin's head spin. Each new year has brought with it changes in the emission control systems on practically every new car, and with each round of changes have come cries of help from the service industry.

Historically, automobile manufacturers have all but ignored the need for mechanic training and service literature outside their dealer network. Repair shop, service station, and do-it-yourself mechanics have been hard-pressed to keep pace with all the new technology and changes under the hood (a good reason for a book like this). The result has been a nightmare of service headaches for mechanics—and a huge increase in repair and drivability complaints by motorists. Mechanics who do not understand how emission control affects engine performance cannot fix today's cars right the first time, the second time, or by trial and error and have no business being under the hood. The systems have become too complex for guesswork.

Prior to the beginnings of emission control in the early 1970s, automotive technology had remained relatively unchanged for the previous 30 or more years. If you could tune a 1939 Ford, you could probably tune a 1969 Ford. There was very little, if any, change in the basic fuel and ignition systems from one year to the next—or from one manufacturer's engines to another's. Emission control changed all that.

With the introduction of emission legislation, every automobile manufacturer started off on its own course of emission control development. Each manufacturer evolved a different approach to meeting the government's standards, and each successive year saw countless changes and "improvements" in ignition, carburetion, and emission control.

Ignition points gave way to pickup modules and black box electronics. Carburetors evolved electronic feedback mixture control and offshoots such as throttle body injection. A host of new gadgets were crammed into the once-uncluttered engine compartments: smog pumps, EGR valves, ported vacuum switches, vacuum amplifiers, vacuum delay valves, engine sensors, and spaghetti-like masses of vacuum hoses.

Today we have such a conglomeration of ignition, carburetion, fuel injection, emissions, and computerized engine control systems that the level of technical expertise required to service such systems is far beyond that of any other blue-collar occupation.

In retrospect, the best approach to clean air would have been for the industry to have worked together in developing common emission control, ignition, and fuel systems, and then to have installed those systems on all vehicles in graduated steps. At the same time, the industry should have set up a program for training mechanics and providing service literature to anyone who needed it. That is what should have been. But in a complicated and competitive world, that type of cooperation is rare. Nevertheless, it is a dead issue now because it is all history.

EARTH DAY AND THE EPA

The late 1960s and early 1970s were a period of great turmoil and social change in this country. People were taking a look at themselves, their society, and their traditional values—and many did not like what they saw: war, poverty, social unjustice, and rampant pollution of our air and water. Like the other problems of that time, pollution was nothing new. The only thing new about it was that it was getting progressively worse.

The mood of the times was such that people wanted to change everything they saw wrong with the world around them. They really believed that they could change things for the better, and so the environmentalist movement was born. People rallied

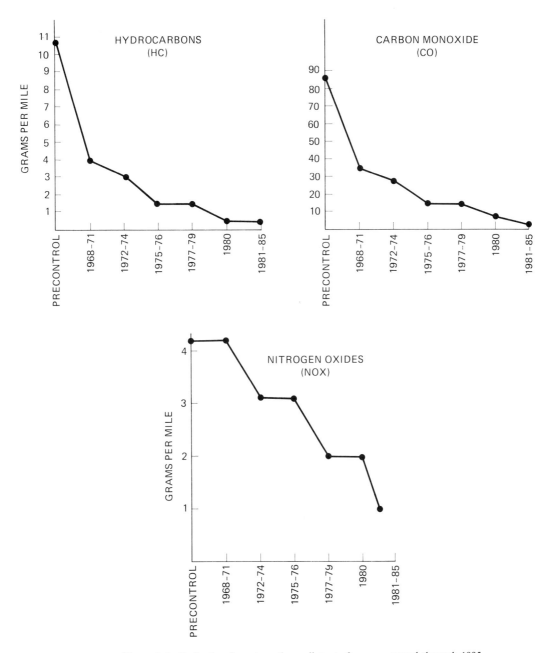

Figure 1-1 Reduction in automotive pollutants from precontrol through 1985.

around the message of Earth Day and supported legislation that would set in motion efforts to clean up the mess we had created. The result was the Clean Air Act of 1970 and the creation of the Environmental Protection Agency.

Today, we still have war, poverty, and social unjustice—but we do have a significantly cleaner environment. Cleaning up the air we breathe met with greater success than some of the other problems of a decade ago because pollution was something technology could solve. Emission control has eliminated 96 percent of the hydrocarbons, 96 percent of the carbon monoxide, and 76 percent of the oxides of nitrogen from automotive exhaust (Fig. 1–1). That is quite an accomplishment for an industry that, 10 years earlier, said it couldn't be done.

EMISSIONS CONTROVERSY

Ever since the creation of the Environmental Protection Agency, (EPA), there has been a controversy over how clean the air should be. The EPA was created to establish clean air standards and then to be the government's watchdog to make sure that those standards were met by industry. However, determining clean air standards is not as easy as it sounds. First, there must be some standard of naturally clean air against which to compare polluted air. Second, the main sources of pollution must be identified—and then controlled.

From the very beginning, the automobile was singled out by the EPA as one of the major sources of air pollution. Scientists were not exactly sure at the time how automobile exhaust formed smog, but they did know that it was a contributing factor. A flurry of scientific activity produced conflicting results, and even today there are those who disagree with many of the basic assumptions that led to strict emission control standards for automobiles. The dissenters say that stationary sources of air pollution, such as factories, steel mills, coal-burning utility plants, refineries, and municipal waste-disposal facilities, are the major polluters—not the automobile. They argue that emission control standards for automobiles should be relaxed—or even repealed—to save fuel and improve drivability. They question the cost-benefits of spending billions of dollars to gain tiny increments in reduced exhaust emissions. Whatever the merits of the conflicting points of view, the fact remains that the EPA established certain air quality standards for automobiles and expects those standards to be met and maintained.

At the time the first standards were set, very few data were available to define actual air quality needs. What's more, emission control technology was in its infancy. The catalytic converter was an unproved promise. Nobody knew whether or not it would work in the real world as it did in a laboratory.

According to those inside the industry, the EPA carefully studied what the auto industry estimated it could achieve by 1980 in terms of reducing exhaust emissions—and then made those estimates law for the 1975 model year, not 1980! This was done to spur the rapid development of new technology, or as one senator put it, "to keep

TABLE 1-1 AUTOMOBILE EXHAUST EMISSIONS REDUCTION[a]

Model year	Hydrocarbons		Carbon monoxide		Nitrogen oxides	
	Grams	Reduction(%)	Grams	Reduction(%)	Grams	Reduction(%)
Pre-control[b]	10.6	—	84.0	—	4.1	—
1968–1971	4.1	62	34.0	60	4.1	0
1972–1974	3.0	72	28.0	67	3.1	24
1975–1976	1.5	86	15.0	82	3.1	24
1977–1979	1.5	86	15.0	82	2.0	51
1980	0.41	96	7.0	92	2.0	51
1981–1985	0.41	96	3.4	96	1.0	76

[a]Federal 49-state standards (grams per mile). Emission controls have drastically reduced automobile exhaust pollution compared to pre-control years. In 1972, EGR was introduced, followed by catalytic converters in 1975 and computerized engine controls in 1980 (California) and 1981.

[b]Motor Vehicle Manufacturers Association 1960 baseline data.

the industry's feet to the fire." The automobile manufacturers screamed and pleaded and used every legal maneuver their lawyers could think of to postpone or reduce the standards—all in vain. The standards dictated a 90 percent reduction in hydrocarbon and carbon monoxide emissions by 1975 and that was it (Fig. 1-1). If the industry could not meet the standards, said the government, they would not be allowed to sell new cars in the 1975 model year. The industry was in an uproar and predictions of doom were proclaimed from industry leaders. But in June 1973, General Motors released a press statement saying that a catalytic converter had been developed that would last 50,000 miles and would meet the EPA emissions requirements. Furthermore, the converter would be installed on all new cars beginning in the 1975 model year (Figs. 1-2 and 1-3).

As the 1975 model year drew nearer, there were dire predictions that the catalytic converter might create more pollution than it eliminated. Because the catalyst caused an oxidation reaction in the hot exhaust gases (thus converting the hydrocarbon and carbon monoxide pollutants into harmless water vapor and carbon dioxide), tiny amounts of sulfur in the exhaust would turn into sulfate (rotten-egg odor). The sulfate, in turn, reacted with the water vapor to form sulfuric acid. Careful tests proved, however, that the amount of sulfates and sulfuric acid produced were so small that there was no significant threat to the environment or public health.

There were also fears that the high temperatures inside the catalytic converter might cause grass fires under cars parked on lawns or near rubbish. A test was conducted for the benefit of the press by the EPA where several cars equipped with catalytic converters and several without were parked in ankle-deep dry grass. The cars were allowed to idle while onlookers waited to see what happened. To the delight of EPA officials, only one grass fire resulted—and that was under a car that was *not* equipped with a catalytic converter.

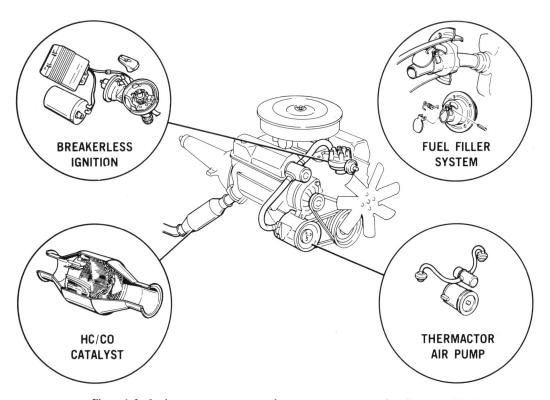

Figure 1–2 In the pre-computer years, these components were primarily responsible for significant reductions in air pollution. (Chevrolet)

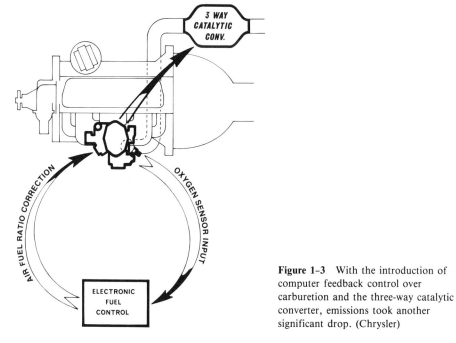

Figure 1–3 With the introduction of computer feedback control over carburetion and the three-way catalytic converter, emissions took another significant drop. (Chrysler)

CARB AND CATF

While the EPA was fighting with Detroit over clean air standards, the state of California decided to take matters into its own hands. For years, Los Angeles had suffered from the worst air pollution in the nation because of its geography and its automotive life-style. Warm air over the Los Angeles basin typically formed an "inversion layer" which trapped pollution underneath. The surrounding mountains prevented the wind from blowing the pollution away, so a huge dome of stagnant, polluted air hung over Los Angeles like a cloud. Californians did not care to wait for the federal emissions standards, so they created the California Air Resources Board (CARB) to implement clean air standards of their own.

CARB led the nation's battle to clean up automotive exhaust by setting standards that were roughly one to two years ahead of the rest of the country. CARB enforced its regulations by banning the sale of any new car that could not meet its emissions tests. As a result, certain engines and drivetrains that were offered in the rest of the country were not available in California because the automobile manufacturers could not meet the standards. To the irk of California consumers, they found that they had to pay higher prices for new cars because of the added pollution equipment.

The situation in California created untold headaches for the car companies because they had to build cars capable of meeting one set of standards in California and another set of standards in the other 49 states. Yet the cloud had a silver lining. Because California's standards were one to two years ahead of the rest of the country, it gave the car companies an opportunity to get the bugs out of their new emission control systems before the equipment was installed on all their cars. But the advantage was slight compared to the manufacturing and service problems created by two sets of emission standards.

To combat CARB, the automobile manufacturers got together with the aftermarket parts manufacturers and repair industry to form the California Automotive Task Force (CATF). CATF's purpose was and still is to oppose whatever CARB deems is in the public interest if it adversely affects the interests of the automotive industry. The result has been some compromises on CARB's part to bring their standards in line with the rest of the 49 states.

THE GREAT PMVI DEBATE

One of the approaches CARB undertook to ensure clean air for the Los Angeles basin was to implement a program of *periodic motor vehicle inspection* (PMVI). The idea behind PMVI, or inspection and maintenance (I/M) programs as they are sometimes called, is to inspect vehicles once a year to make sure that the emission control equipment is in good working condition. This is accomplished by subjecting the vehicle to a tailpipe emissions test and an underhood inspection. The tailpipe test certifies that the vehicle's emissions are within the limits set by law. The underhood inspection verifies that the pollution control equipment has not been disconnected or tampered with.

Although the subject of vehicle emissions inspections is surrounded with emotion and controversy, the fact remains that both CARB and the EPA are pushing PMVI as a means of ensuring continued air quality. The EPA wants every major metropolitan area of the country that fails to meet its air quality standards to implement a PMVI program. Many have and most eventually will despite all the politics that surround the issue.

Consider the following example: According to the EPA, New York is second only to Los Angeles for having the nation's foulest air. Each day, 4 million cars and light trucks pour 1400 tons of carbon monoxide and 600 tons of hydrocarbons into the atmosphere over nine metropolitan New York counties. That is double the EPA's maximum allowable levels. Cars account for about 90 percent of the total carbon monoxide problem and half of hydrocarbon emissions in the metropolitan area.

For years, the federal government pressured New York to clean up its air. Specifically, the Clean Air Act amendments of 1977 required the state to get hustling on a long-range program to improve air quality in the nine-county downstate area. The EPA mandated goal: to eliminate 25 percent of CO and HC pollutants over a six-year period. The method: mandatory annual inspection and maintenance of exhaust emissions for all registered vehicles 8500 lb or less gross weight. The alternative: to lose over half a billion dollars in federal funds as a consequence of not meeting the government's clean air standards. New York had no choice but to implement a "decentralized" PMVI program for the nine-county area in January 1981. (By the end of 1984, 24 states had emission inspection laws in effect.)

New York was the first state to implement a decentralized inspection program. Unlike "centralized" inspection programs, service stations and garages authorized by the state conduct the vehicle inspections. In a centralized program, the vehicles are inspected by state-run inspection centers. If a vehicle fails, it can be taken to the motorist's local repair shop, service station, or dealership for repairs. The car is then brought back for reinspection to make sure that the problem was corrected.

The main criticism of centralized inspection programs is inconvenience. Motorists often find that they have to wait in long lines to have their vehicles inspected. And if they fail, they can look forward to a trip back for reinspection.

The decentralized approach, such as that used by New York, allows private repair outlets to inspect, repair, and reinspect their customers' vehicles. The criticism of this approach is that it opens the door for fraud on the part of the repair establishments because they stand to profit on any vehicle that fails to pass an emissions test.

To minimize the possibility of fraud against motorists as much as possible, New York required that all the official inspection stations be equipped with the same type of exhaust analyzing equipment; furthermore, that the analyzers be equipped with printouts so that the consumer can see the actual readings for the vehicle. Each exhaust test is also recorded on a magnetic tape so that the state can monitor the test results.

WHO PAYS FOR CLEAN AIR?

Everybody's for clean air but few want to shoulder the burden of paying for it. Thus CARB and EPA promoted the concept of the emissions warranty. To reduce motorists' out-of-pocket expenses for maintaining the pollution control equipment on their vehicles, the government decided that it would be politically smart to shift the burden of clean air to the automobile manufacturer in the form of an extended emissions warranty. The first such regulations required a two-year/24,000-mile warranty on all the major pollution control equipment on new cars. If anything went wrong with the equipment in that period, the motorist would return the car to the dealer for free repairs.

Although the two-year/24,000-mile emissions warranty regulations met with industry resistance, both sides eventually agreed the terms were not unreasonable. But then CARB decided to push the issue even further. CARB wanted to require a five-year/50,000-mile emissions warranty that in effect would have covered practically everything on the car—anything that even remotely affected emissions would be covered by the warranty. This included spark plugs, the carburetor, the entire ignition system, filters, and even the tires. CATF and the industry said this was going too far. Service stations and independent repair shops argued that such a warranty would put them out of business, since new-car dealerships would get all the business during the first five years of a vehicle's life. Aftermarket repair organizations argued that if their members went broke, the new-car dealerships would not be able to handle all the business, simply because there weren't enough of them.

The issue went around and around, with the eventual result being that CARB and EPA required a five-year/50,000-mile warranty only on major emission control devices—not everything under the hood. The issue still is not dead, and further changes may come depending on the mood of the people and government. There is talk of a 10-year/100,000-mile emissions warranty and talk of doing away with all emissions warranties altogether.

The next great emission battle will be over diesel engine emissions. Diesel engines are more efficient than gasoline engines—but also more dirty. With the increasing popularity of automotive diesel engines, it seems likely that the diesel will eventually have to be equipped with a particulate trap in its exhaust system together with other emission control devices.

THE FUTURE OF EMISSION CONTROL

We seem to have reached a plateau in automotive exhaust emissions. The catalytic converter and computerized engine controls have allowed us to cut emissions about as far as they can be realistically reduced. Any further improvements would probably not be cost-effective. The EPA apparently agrees as there are no further planned reductions in exhaust emissions beyond the levels established back in 1981.

As mentioned earlier, vehicle emissions inspections will become a way of life for most motorists in large metropolitan areas. The cost of emission control maintenance will be borne primarily by consumers—in the form of higher new-car prices to cover the cost of extended emissions warranties. For older cars, the costs will come directly from the consumer's pocket at the time of repair.

It is very unlikely that today's emissions regulations will ever be rescinded. We have come too far for that to happen. The EPA says that general air quality has improved greatly and that it will continue to improve as long as we continue to maintain and enforce emission control. There is the possibility that today's tough standards may be modified or reduced somewhat if it can be demonstrated that the benefits would outweigh the disadvantages. The United States has the toughest emissions standards in the world—maybe too tough. Industry experts have said that if we reduce our standards to match those of many other industrial countries, new-car fuel economy would be improved an average of 15 percent and the cost reduced by an average of $400 per vehicle. They argue that the slight increase in pollution would have a negligible effect on public health.

Time will tell who will eventually win the emissions battle. Hopefully, we all will win. In the meantime, you, the mechanic, will be responsible for keeping today's emission-controlled engines running and solving the drivability problems that result when something goes wrong.

With that in mind, let's move to the next chapter.

2

Basic Pollutants

Many people tend to think of automotive air pollution as being strictly exhaust emissions, but cars can actually pollute the atmosphere three ways:

1. Gasoline vapors from the fuel tank and carburetor
2. The combustion by-products and vapors from the engine's crankcase
3. The exhaust gases that come out the tailpipe

EVAPORATIVE EMISSIONS

Gasoline fuel vapors, or *evaporative emissions,* as they are called, include a variety of hydrocarbons (HC). This is because gasoline itself is a blend of many different hydrocarbons. The makeup of any given gallon of gasoline depends on the grade of crude oil from which it was refined, the refining process it underwent, and the additives that were put in to improve the performance characteristics of the fuel. Because of such factors, the nature of gasoline can vary considerably from one area of the country to another, or even from one brand to another in the same town.

The lighter elements in gasoline evaporate easily, especially in warm weather. These include aldehydes, aromatics, olefins, and higher paraffins. These substances can react with air and sunlight (called a *photochemical* reaction) to form smog. Aldehydes are often called *instant smog* because they can form smog without undergoing photochemical changes.

Evaporative emissions account for about 20 percent of automotive emissions. When you consider the fact that a parked car can pollute the air with hydrocarbon emissions even though the engine is not running, controlling evaporative emissions becomes an important consideration.

Evaporative emissions are eliminated by sealing off the fuel system from the atmosphere. This prevents the gasoline vapors from evaporating out of the fuel tank or carburetor bowl. Vent lines from the fuel tank and carburetor bowl route vapors to a charcoal canister, where they are stored until the engine is started. The vapors are then drawn into the intake manifold and burned. Evaporative emission control was required on all cars sold in California in 1970, and on all other cars in the United States in 1971. Evaporative emission control systems are explained in Chapter 4.

CRANKCASE EMISSIONS

When the air/fuel mixture explodes inside the combustion chamber, the tremendous pressures cause some of the gases to blow by the piston rings and enter the engine's crankcase. Since the blowby gases contain water vapor and unburned gasoline, they will dilute the oil if allowed to accumulate in the crankcase. This, in turn, would promote sludge formation, reduce the oil's ability to lubricate, and shorten engine life. So to prevent such calamities from happening, some means of venting the crankcase must be provided.

Before the days of emission control, crankcase ventilation was accomplished with an open oil breather cap and a road-draft tube. Crude and not very efficient, the design allowed the blowby gases to escape out the road-draft tube and into the atmosphere. The water vapor causes no problems, but the hydrocarbons contribute to the formation of smog.

Crankcase emissions, which can account for 20 percent of automotive emissions, are eliminated by sealing the crankcase and routing the blowby gases back into the intake manifold. The system that does this is known as *positive crankcase ventilation* (PCV). First used in California in 1961, PCV became required equipment on all cars in 1963. Not only does PCV eliminate crankcase emissions, it also greatly improves crankcase ventilation by using engine vacuum to suck the blowby gases out of the crankcase. So in addition to reducing air pollution, PCV prolongs oil life and ultimately engine life. The PCV system is explained in Chapter 3.

EXHAUST EMISSIONS

Exhaust emissions are the most difficult to control because there are so many variables. The most important factors are the air/fuel ratio, ignition timing and advance, combustion chamber design, combustion temperature, and fuel composition. Other factors that infuence what comes out the tailpipe include camshaft timing, intake manifold design, engine compression, and external factors such as humidity, temperature, and barometric pressure. These are explained in greater detail in Chapters 9 to 11.

There are two basic reasons for exhaust emissions. One is incomplete combustion. The other is the presence of unwanted substances in the combustion process.

When gasoline is burned inside an engine, combustion is never totally complete. There is always a tiny amount of fuel that fails to burn or is only partly burned. To understand why this is so, we need to look at the basic chemistry of combustion.

Gasoline, as described earlier, is a blend of various hydrocarbons. A hydrocarbon is any substance whose molecules contain hydrogen and carbon atoms. The molecules in gasoline consist of long chains of hydrogen and carbon atoms.

When gasoline is drawn through the carburetor venturis or sprayed out of a fuel injector, it is broken up into tiny droplets. To burn, the gasoline droplets must be mixed with oxygen in the right proportion. Since the air we breathe contains 16 to 20 percent oxygen, it takes a lot of air to provide enough oxygen for combustion. The *ideal* or *stoichiometric* air/fuel ratio for gasoline is 14.7:1. That means 14.7 parts of air for every 1 part of gasoline.

The air/fuel mixture is drawn into the combustion chamber, compressed, and ignited by a spark. Now if combustion were totally complete and the air and fuel were mixed in the correct proportions, all the oxygen would combine with all the gasoline to produce heat energy, water vapor, and carbon dioxide:

$$O_2 + HC = H_2O + CO_2$$

Since water vapor and carbon dioxide are harmless by-products of combustion, we would have no air pollution problems *if* combustion inside a real engine were this simple. But it is not. For one thing, the relatively cool surfaces of the piston, cylinder walls, and cylinder head have a quenching effect on the tiny droplets of fuel. Some of the fuel will cling to the cool metal surfaces and not burn. This results in unburned hydrocarbons in the exhaust.

Intake manifold design can also interfere with combustion by allowing the tiny droplets of fuel to separate from the airflow. This causes an uneven mixing of the air and fuel, which hinders complete combustion.

Sometimes the air/fuel ratio is not right. If the mixture is too rich (too much fuel, not enough oxygen), there will not be enough oxygen to burn all the fuel completely. The result is incomplete combustion and the formation of carbon monoxide (CO) in the exhaust:

$$\text{insufficient } O_2 + HC = H_2O + CO$$

If the air/fuel mixture is really short on oxygen, the result will be visible particles of carbon soot in the exhaust:

$$\text{insufficient } O_2 + HC = H_2O + C$$

Excessive oxygen can also create emission problems. If the air/fuel mixture is too lean (too much oxygen, not enough fuel), a condition known as *lean misfire* will occur where the mixture is too lean to ignite. This will allow the gasoline to pass through the engine completely unburned and increase hydrocarbon emissions out the tailpipe.

High combustion temperatures inside the engine also complicate the combustion process. Air is 80 percent nitrogen. Normally, nitrogen is inert and does not do much of anything. But at temperatures above 2500 °F, nitrogen and oxygen combine to form oxides of nitrogen. The abbreviation NO_x is used to describe the various oxides of nitrogen because any one of several different compounds can be formed. The higher the combustion temperature, the greater the tendency to form NO_x. In an uncontrolled engine, combustion temperatures can easily exceed 2500 °F, so some means of lowering the temperatures must be used to minimize the formation of NO_x. This system is known as *exhaust gas recirculation* (EGR). The EGR system is explained in Chapter 8.

What else can contribute to exhaust emissions? Oil seeping past the rings or valve guides will increase hydrocarbon emissions. Lead, which is used to boost the octane rating of regular gasoline, passes right through the combustion process to produce tiny amounts of lead in the exhaust. Sulfur, which is present in varying amounts in crude oil, combines with oxygen during combustion to form SO_2—the rotten-egg odor characteristic of some catalytic-converter-equipped cars. When SO_2 combines with water vapor in the exhaust, the result is sulfuric acid. Table 2–1 lists the basic emissions control devices and the pollutants that each is designed to reduce or eliminate.

TABLE 2-1 BASIC EMISSION CONTROL DEVICES AND THE POLLUTANTS EACH AFFECTS.

Method	Function
Heated air intake system	Reduces HC in the exhaust
Positive crankcase ventilation (PCV)	Reduces crankcase HC
Carburetor air/fuel ratio	Reduces HC and CO in exhaust
Ignition timing advance	Reduces HC, CO, and NO_x in exhaust
Air pump	Works with the catalytic converter to reduce HC and CO in the exhaust
Exhaust gas recirculation (EGR)	Reduces NO_x in exhaust
Evaporative control system	Prevents HC and other vapors from escaping from the fuel tank or carburetor
Catalytic converter	Two-way catalyst reduces HC and CO in the exhaust
	Three-way catalyst reduces HC, CO, and NO_x in the exhaust
Unleaded gasoline	Eliminates lead pollution in exhaust

Now let's take a closer look at some of the basic pollutants to see why they are considered harmful to human beings and/or the environment (see also Table 2–2).

TABLE 2-2 POLLUTANTS STANDARDS INDEX[a]

Index value	Air quality level	Micrograms per cubic meter				Milligrams per cubic meter:	Health effect description	General health effects	Cautionary statements
		Total suspended particulates (24-hour)	Sulfur dioxide (24-hour)	Ozone (1-hour)	Nitrous oxide (1-hour)	Carbon monoxide (8-hour)			
500	Significant harm	1000	2620	1200	3750	57.5		Premature death of ill and elderly. Healthy people will experience adverse symptoms that affect their normal activity.	All persons should remain indoors, keeping windows and doors closed. All persons should minimize physical exertion and avoid traffic.
400	Emergency	875	2100	1000	3000	46.0	Hazardous	Premature onset of certain diseases in addition to significant aggravation of symptoms and decreased exercise tolerance in healthy persons.	Elderly and persons with existing diseases should stay indoors and avoid physical exertion. General population should avoid outdoor activity.
300	Warning	625	1600	800	2260	34.0	Very unhealthful	Significant aggravation of symptoms and decreased exercise tolerance in persons with heart or lung disease, with widespread symptoms in the healthy population.	Elderly and persons with existing heart or lung disease should stay indoors and reduce physical activity.
200	Alert	375	800	400[b]	1130	17.0	Unhealthful	Mild aggravation of symptoms in susceptible persons, with irritation symptoms in the healthy population.	Persons with existing heart or respiratory ailments should reduce physical exertion and outdoor activity.
100	NAAQS[c]	260	365[e]	160	d	10.0	Moderate		
50	50% of NAAQS	75[e]	80[e]	80	d	5.0	Good		
0		0	0	0	0	0			

[a]Pollutants standards index for air quality as issued by the Environmental Protection Agency.
[b]400 micrograms per cubic meter used instead of the ozone alert level of 200 micrograms per cubic meter.
[c]National Ambient Air Quality Standard.
[d]No index values reported at concentration levels below those specified by "Alert Level" criteria.
[e]Annual Primary NAAQS.

Source: Environmental Protection Agency.

HYDROCARBONS

With respect to automotive emissions, hydrocarbons include gasoline vapors from the fuel system and crankcase, and unburned gasoline and oil vapors in the exhaust. Hydrocarbons (HC) are measured in parts per million (ppm) using an infrared exhaust analyzer.

Hydrocarbons react with sunlight in the atmosphere to form smog. Smog, in turn, can irritate the eyes and nose, cause breathing problems in the elderly or those with lung problems, and cause headaches, dizziness, and a whole range of symptoms in persons who are sensitive to polluted air. There is also evidence which suggests that certain hydrocarbons may cause cancer.

PARTICULATES

Particulates are the tiny particles of carbon soot that result from incomplete combustion and excessively rich air/fuel ratios. Diesel engines are notorious for high-particulate emissions. Particulate emissions are measured in the laboratory and are specified in grams per mile.

Scientists estimate that as much as 100 tons of particulates fall to the ground or are breathed in by the public in most large American cities every day. Many particulates are smaller than 1 micron in diameter (39 millionths of an inch). Particles this small are called *aerosols* because they tend to remain suspended in air rather than settle to the ground. Such particles pose a significant health hazard because they are too small to be filtered out by the mucous membranes in the nose and windpipe. They penetrate deep into the lungs and remain there, accumulating with each passing year. This can increase the odds of developing emphysema and cancer, especially if a person smokes.

CARBON MONOXIDE

Carbon monoxide is an odorless, colorless, *deadly* gas. It has earned a reputation for being a silent killer. Carbon monoxide (CO) results from incomplete combustion where insufficient oxygen is available to convert the carbon into carbon dioxide. The richer the air/fuel ratio, the greater will be the amount of carbon monoxide produced in the exhaust. An infrared exhaust analyzer is used to measure the percent of carbon monoxide in the exhaust.

Unlike some of the other pollutants, carbon monoxide is the only one that can be deadly. It can kill in less than 10 minutes if you are inside a small closed garage with a motor running. It does so by displacing the oxygen in your bloodstream. A molecule of carbon monoxide has 210 times the affinity with joining red blood cells as does oxygen. Breathing it in reduces the amount of oxygen that reaches your brain. This starves the brain for oxygen and rapidly leads to unconsciousness and death. Breathing lesser concentrations of carbon monoxide can also be hazardous to your health. When the CO concentration in a person's blood reaches 2 to 5 percent, vision begins to blur and reaction time starts to drop. It can also cause headaches, dizziness, chest pains, and breathing difficulty. Carbon monoxide concentrations of 20 ppm are not uncommon in heavy traffic.

OXIDES OF NITROGEN

Oxides of nitrogen (NO_x) are formed when oxygen and nitrogen combine inside the combustion chamber. This process begins when combustion temperatures exceed 2500°F and increases with rising temperature. NO_x cannot be read with an infrared exhaust analyzer. It requires special costly equipment that is normally only used in a laboratory.

NO_x is a nasty pollutant because it reacts with sunshine and hydrocarbons to produce smog. It also causes the dirty brown haze associated with smog. It has an odor that becomes noticeable in concentrations as small as 1 to 3 ppm. When levels reach 5 to 10 ppm, NO_x causes eye and nose irritation in some people. Prolonged exposure to 10 to 40 ppm can trigger emphysema attacks. Higher concentrations can cause bronchitis and other lung disorders.

Most of the NO_x that comes out the tailpipe is in the form of nitric oxide (NO), a colorless posionous gas. It then combines with oxygen in the atmosphere to form nitrogen dioxide (NO_2), which is the brownish version of the gas. NO_x also undergoes other reactions in the atmosphere to produce ozone.

Ozone is probably one of the most toxic and dangerous air pollutants known. Ozone is formed when an extra atom of oxygen attaches itself to the normal oxygen molecule:

$$O_2 + O = O_3 \quad \text{(ozone)}$$

The extra oxygen atom causes the molecule to be very toxic to other materials. It irritates the eyes and lungs. It causes a variety of symptoms ranging from coughing, headaches, choking, and a feeling of weariness in concentrations as low as 0.5 ppm. Ozone attacks rubber products and is toxic to many types of plants and microbes. In fact, industry sometimes uses ozone under controlled conditions to disinfect certain products. The maximum concentration of exposure over an 8-hour period to ozone is considered by most medical experts to be 0.1 ppm.

LEAD

Lead pollution is not a problem with cars that burn unleaded gasoline. But for vehicles that do use leaded regular, lead pollution is a problem. Lead is used in gasoline for two reasons. One is to raise the octane rating of the fuel. This allows a refiner to take a lower grade of gasoline and boost its performance to acceptable levels by adding tetraethyl lead. Federal regulations limit the amount of lead that can be added by refiners. But many secondary refiners or blenders add up to five times the maximum permissible amount of lead to marginal gasoline to make the product salable. Lead also has a lubricating effect on exhaust valves which helps to prolong their life.

Now for the bad news. If leaded gasoline is used in a car with a catalytic converter, the lead fouls the catalyst and renders it useless. This, in turn, increases HC and CO emissions.

Lead is considered a serious threat to the environment because it literally lasts forever. When ingested into the body either by breathing in lead-polluted air or by drinking water from lead-polluted wells or by eating plants contaminated by lead dust or lead-polluted soil, the lead accumulates in the body and eventually produces lead poisoning. There is evidence to suggest that lead from automobile exhaust in metropolitan areas adversely affects the learning abilities of schoolchildren. Lead is absorbed into nerve cells in the brain, where it inhibits normal functions. Tests have shown that schoolchildren who live near areas of traffic congestion score lower on intelligence tests and are more prone to behavior problems than those who live away from lead-polluted environments.

The Environmental Protection Agency is gradually phasing out lead, with a complete ban on leaded fuel planned by the early 1990s.

SULFUR

All crude oil contains a certain amount of sulfur. Crude oil from the western United States contains much more sulfur than do eastern crude oils. When the crude is refined to make gasoline, some of the sulfur remains in the fuel. The concentrations are small, but are large enough to cause pollution problems.

When gasoline containing sulfur is burned, the sulfur combines with oxygen to form sulfur oxides. When sulfur oxide concentrations reach 8 to 12 ppm, most people's eyes will water. Such levels also cause coughing, breathing difficulty, and can increase the likelihood of heart disease over time. Sulfur dioxide (SO_2) produces the familiar rotten-egg odor. This can sometimes be a problem with catalytic-converter-equipped cars that burn gasoline laden with high sulfur concentrations.

Sulfur oxides also have another bad effect on the environment. They react with water vapor to form sulfuric acid. The result is acid rain. This is what accounts for the deterioration and erosion of buildings and statues exposed to polluted air.

Acid rain is causing long-term changes in our environment. The Adirondacks is a pristine wilderness area in upstate New York. There are no factories and few roads. Yet acid rain caused by pollution sources hundreds of miles upwind is destroying many lakes in the area. Over 200 lakes that once provided excellent fishing are now devoid of marine life because acid rain has made the lake water toxic to fish. Hundreds of other lakes in the region are threatened by the same fate. The lakes are dying because they lack any natural means of countering the cumulative effects of acid rain. In lakes with limestone deposits, the lime neutralizes the acid and prevents the buildup of dangerous concentrations of acid. But the lakes in the Adirondacks lack limestone, so their days appear numbered.

EMISSION CONTROL

The worst pollutants, HC, CO, and NO_x, can be controlled by today's emission control systems. If maintained properly, these systems can keep automotive air pollution to a minimum. Use of unleaded, low-sulfur gasoline can also eliminate lead and sulfur emissions. But the best emission control systems can be rendered useless when someone tampers or disconnects to "improve" performance. Using leaded gasoline to save a few cents at the pump also destroys the benefits of emission control.

Any one vehicle by itself will not make or break the environment. But when you think of the 150 million vehicles on the road added together, you begin to realize the significant impact that emission control has on our nation's air quality. The Environmental Protection Agency says that air quality bottomed out in the mid-1970s, and has been slowly improving ever since. We still have a long way to go. But in many areas, the air is now as clean as it was in 1960. Thanks to emission control, the automobile is becoming less and less of a polluter (see Tables 1–1 and 1–2). The emphasis is now on stationary sources of pollution—factories and utilities that are burning high-sulfur-content coal and oil.

3

Positive Crankcase

Ventilation:

Airing Out the Engine

During the 1950s, our society as a whole began to be concerned with what our industrial way of life was doing to our environment. Books appeared on the subject of pollution and jokes about the smog that hung like a shroud over Los Angeles were rife. It was beginning to become apparent to a great number of people that we were mortgaging our future and that of our progeny, and, as difficult and expensive as it might be, corrective steps were going to have to be taken to avert catastrophe.

The automakers, being citizens of the world too, were concerned because the vehicles they produced were mobile machines that spread pollution wherever they went. Cars and trucks were committing some of the worst offenses against clean air.

A GOOD START

Because of this concern and worry about the possibility of government intervention in the form of far-reaching laws, the car makers voluntarily added positive crankcase ventilation (that is what PCV stands for, not "pollution control valve" as some television advertisements of the past would have us believe) systems to all engines sold in this country in 1963. It was a good start.

This efficient, sure way of airing out the crankcase is made necessary because even the most precise, well-engineered set of piston rings possible cannot seat against the cylinder walls perfectly. There are numerous reasons for this: The pressure is too great; there is a great deal of movement involved; rings must necessarily have gaps; there has to be clearance between the rings and the piston groove lands, which allows a certain amount of flex and twist; pistons rock a little in their bores; some fuel stays in the pores of the cylinder wall surfaces; and so on.

This inability of the Otto cycle piston engine to contain all of the fuel and burning gases within the combustion chamber and cylinder results in *blowby,* the passage of raw fuel and the products of combustion into the crankcase.

THE OLD WAY

In the "B.C." (Before Controls) era, the crankcase was simply vented to the atmosphere to allow these gases to escape and so keep pressure from building up inside the engine (Fig. 3–1). There was a road-draft tube that ran down under the chassis at an angle that produced a small amount of vacuum as the vehicle traveled forward. Fresh air was drawn through a mesh filter in the oil filler cap, circulated around inside the crankcase, and exited through the road-draft tube carrying blowby with it.

The increased pressure found in the engine compartment that was caused by air being forced through the grill as the car rolled forward, and slightly enhanced by the fan, helped this circulation in a small degree by adding to the pressure differential between the air inlet and the end of the exit tube. Also, in many cases, the outlet was positioned in the block at a point that took advantage of the pumping action set up by the rotation of the crankshaft.

The gases were mostly HC (hydrocarbons) from the incomplete combustion of gasoline and from hot lubricating oil. With a car idling, one could remove the oil filler cap and observe these fumes escaping; they were also visible at the road-draft tube. If the engine was in poor condition internally, this puffing was very noticeable—and serious.

This was the most obvious place to start a campaign against air pollution—and, as it turned out, the least troublesome.

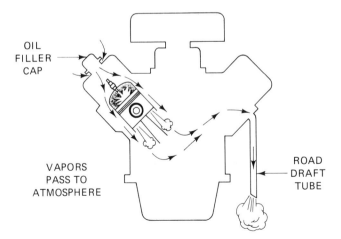

OIL
FILLER
CAP

VAPORS
PASS TO
ATMOSPHERE

ROAD
DRAFT
TUBE

Figure 3–1 Before PCV, blow-by was allowed to escape into the atmosphere through a road draft tube. (Chrysler)

The technology involved was not entirely new. For many years before automotive pollution became a widespread concern, some military and commercial vehicles employed the basic principle of positively airing out the crankcase by means of engine vacuum in order to reduce the degree of internal engine contamination.

The PCV system simply uses the vacuum an engine naturally produces to draw fresh air through the crankcase and into the intake manifold from whence it goes to the cylinders to be burned. Not only did it eliminate about 20 percent of the total amount of emissions a car generates, it also kept the internals a great deal cleaner, which allowed oil change intervals to be extended without risking the harmful buildup of sludge and varnish—that is, provided that the system is working properly.

OPEN TYPE

A typical early PCV setup comprised a hose that routed blowby from a rocker cover into a spacer plate between the carburetor and the intake manifold, and an uncomplicated valve that regulated flow (Fig. 3-2). This was known as the *open system* because fresh air was still admitted through a mesh filter in the oil filler cap. If blowby exceeded the capacity of the valve, say during full-throttle operation, the excess gases could still get into our fragile atmosphere by backing up through the filler cap.

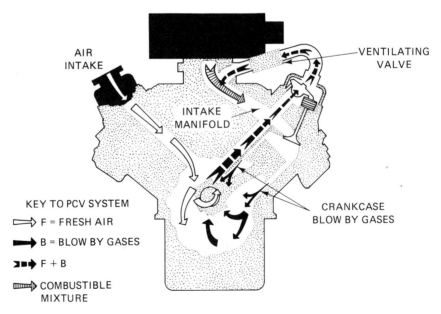

AIR INTAKE

VENTILATING VALVE

INTAKE MANIFOLD

KEY TO PCV SYSTEM

⟹ F = FRESH AIR

➡ B = BLOW BY GASES

⟹ F + B

⟹ COMBUSTIBLE MIXTURE

CRANKCASE BLOW BY GASES

Figure 3-2 The open PCV system was installed on all domestic cars in 1963. It eliminated crankcase emissions except under conditions where the capacity of the valve was exceeded. Then, some blow-by escaped through the air intake. (Chevrolet)

CLOSED TYPE

To eliminate this possibility, the *closed system* was introduced in 1968 (Fig. 3–3). This was a very simple modification consisting of a second hose that ran from the rocker arm cover to the air cleaner housing and terminated in a mesh filter (Fig. 3–4). The oil filler cap no longer was vented, but sealed instead.

In normal operation, the closed system picks up fresh air from inside the air cleaner in the same way the open system did at the oil filler cap. But whenever there was more blowby than could be handled by the valve, the excess backed up through the hose into the air cleaner, where there is always a partial vacuum if the engine is running, so it was drawn into the carburetor along with intake air and there was no chance of it escaping into the atmosphere.

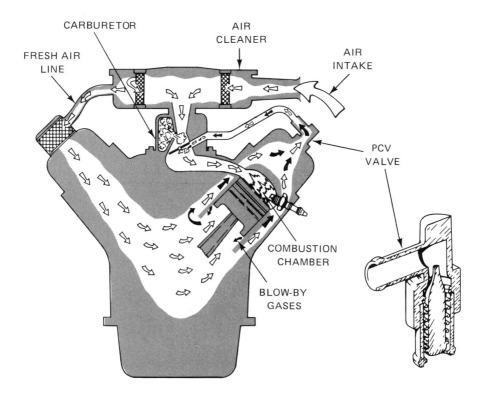

Figure 3-3 The closed PCV system appeared in 1968. Crankcase intake was at the air cleaner so if the valve's capacity was exceeded, blow-by backed up into the intake where, since there's vacuum in the air cleaner whenever the engine's running, it was drawn into the carburetor. (Chrysler)

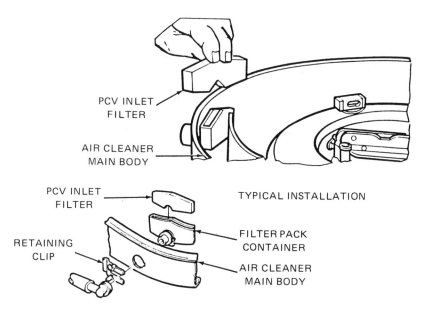

PCV INLET
FILTER

AIR CLEANER
MAIN BODY

PCV INLET
FILTER

TYPICAL INSTALLATION

RETAINING
CLIP

FILTER PACK
CONTAINER

AIR CLEANER
MAIN BODY

Figure 3-4   Closed PCV systems get their supply of fresh air through an inlet filter in the air cleaner. It needs periodic attention, especially if blow-by often exceeds the valve's capacity. (Chrysler)

THE VALVE ITSELF

The PCV valve itself, while fairly simple, is still widely misunderstood, so a discussion of its working principles is important. To begin with, any garden-variety valve consists of a housing [one end usually plugs into a grommet in the rocker cover (Fig. 3-5) and the other end connects to a hose that goes to the spacer under the carburetor]

Figure 3-5   In most specimens, the PCV valve is mounted in a grommet in the rocker or cam cover.

with a movable plunger inside, and a light coil spring that bears against the plunger. The housing may be made of steel or plastic.

One of the characteristics of the gasoline piston engine is that it produces more intake manifold vacuum when idling than when accelerating or cruising, which is simply a function of the amount the throttle plates are open and hence how much atmospheric pressure is allowed to enter the engine. The PCV valve takes advantage of this situation in order to meter the flow of blowby.

At idle, strong vacuum forces the plunger against the outlet port, compressing the spring (Fig. 3–6). The end of the plunger and the opening of the port are made in such a way that in this position only a small amount of PCV flow is permitted. This keeps the system from interfering with the idle mixture any more than is necessary. The flow is sufficient to accommodate the relatively small amount of blowby an engine produces at low rpm rates.

As the throttle plates are opened and the rpm rate increases, vacuum is reduced, and the spring begins to overcome it, forcing the end of the plunger away from its

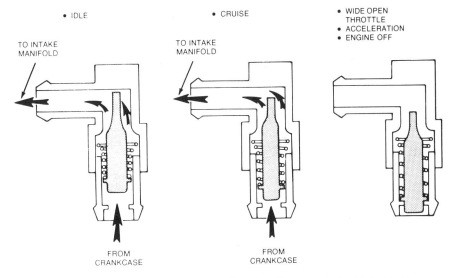

Figure 3–6 PCV valves are designed so they allow less flow during high vacuum conditions (as at idle) than they do when vacuum drops (as when cruising) because more blow-by is generated at higher rpm. With the engine off, at wide open throttle, or during an intake manifold backfire, the spring holds the valve closed. (Chrysler)

seat and allowing a larger volume of waste gases to pass. This is a progression with flow increasing with rpm so that the PCV system keeps up as the engine produces more blowby. A typical valve will allow 1 to 3 cubic feet per minute (cfm) of flow at idle, and 3 to 6 cfm at 6 in. Hg of vacuum.

If there should be a backfire in the carburetor and intake manifold, the pressure shock waves slam the plunger momentarily against a seat in the other end of the housing, shutting the valve off completely and eliminating the possibility of flammable fumes igniting in the crankcase.

IMPORTS WITH NO VALVE

Some imported vehicles do not have a PCV valve. Instead, ventilation of the crankcase is accomplished by means of a one- or two-stage orifice setup. Honda's Dual Return System is an example of this (Fig. 3–7). A liquid-vapor separator is built into the top of the camshaft cover, and this is connected by a hose to a condensation chamber at the bottom of the air cleaner. Another hose runs from the condensation chamber down to a fixed orifice vapor passage connected to the carburetor insulator that is mounted on the intake manifold.

During idle and part throttle operation when vacuum is high, blowby flows from the liquid-vapor separator to the condensation chamber, then down through the fixed orifice and into the intake stream. The fixed orifice is calibrated so that it does not cause idle mixture problems by allowing too much flow.

At throttle positions approaching wide open, vacuum in the air-cleaner housing increases and that in the intake manifold decreases. Therefore, blowby leaves the liquid-vapor separator, enters the condensation chamber, and leaves through

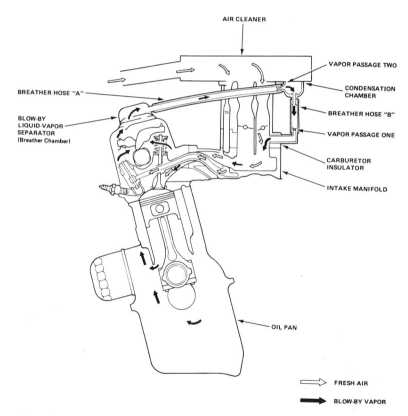

Figure 3–7 This Honda uses two calibrated orifices instead of a PCV valve to regulate the flow of blow-by. No fresh air is admitted. (Honda)

the condensation chamber's upper passage. This exits inside the circumference of the air filter element itself so that the vapors do not soak it and clog it up. Finally, the blowby is drawn into the carburetor throat. Very little enters the intake stream through the fixed orifice during this low-vacuum mode.

Although it may appear that there is nothing to go wrong with this system since no valve is used, there is always the possibility of the orifice becoming plugged, and that would interfere with the idle mixture and cause all the blowby to enter the carburetor through the air cleaner.

ONE RETROFIT SETUP

Another type of system that prevents blowby from escaping uses the pressure in the crankcase and a variable orifice valve. This has been used primarily as an add-on, retrofit setup on pre-1963 cars. It consists of a sealed oil filler cap and a valve that is spliced into a hose between the rocker cover and the intake manifold. The valve remains closed except for a small passage when the engine is idling or when blowby pressure is low. When the engine is speeded up and blowby pressure increases, a spring forces the valve off its seat to allow a greater amount of flow.

NO FRESH AIR

As you can see, there is no provision in these systems for admitting fresh air into the crankcase, so the flow is limited to the amount of blowby that the engine produces. Some imports use a more basic setup in which there is simply one hose that runs from the top of the camshaft or rocker cover to the air cleaner housing with no valve or calibrated orifice in between. The fact that there is pressure inside the engine and a partial vacuum in the air cleaner makes the vapors move into the intake stream. This has the disadvantage of wetting the filter element. Also, when the weather is cold, the water vapor present in the blowby can cause carburetor icing. Although it would seem logical that true crankcase ventilation systems, where fresh air passes through the inside of the engine and carries blowby through a PCV valve into the intake manifold, would tend to keep the internal parts and surfaces cleaner than those arrangements that do not admit fresh air, this is not necessarily the case in the real world.

PROBLEMS AND TESTS

A malfunctioning crankcase ventilation system can cause several problems, the most serious being the accumulation of sludge and varnish inside the engine and, if there is a pressure buildup, leaking gaskets and seals. Other troubles include rough idling or stalling if the PCV valve is stuck open, and the soaking of the air filter element if there is a blockage somewhere. So it is extremely important that the system be kept in good working order.

A common test of a typical system is to remove the PCV valve from its grommet in the rocker cover and shake it to make sure it rattles. This will let you know whether the valve is clogged or jammed, but it cannot tell you if the spring has broken or lost its tension.

Another check is to feel for strong vacuum at the end of the valve while the engine is idling. If there is none, the valve, hose, or passages are blocked. Even if vacuum is present, you still will not know if the valve's spring is weak or broken.

Testers are available that attach to the oil filler hole and measure the vacuum or airflow the system produces (Fig. 3–8). Follow the equipment maker's instructions.

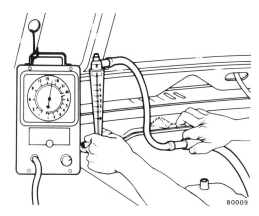

Figure 3-8 A set-up such as this can be used to check the volume of PCV flow against engine vacuum. (American Motors Corp.)

REGULAR RETIREMENT

A PCV valve is an inexpensive part, especially when you consider the value of the engine it is largely responsible for protecting. So it is certainly justifiable to replace it at the intervals the manufacturer recommends, which range from every 12,000 to every 30,000 miles. Of course, practically this becomes a matter of installing a new valve at each tune-up.

4

Evaporative Emission Control

One of the qualities that makes gasoline such a desirable motor fuel is that it evaporates easily. Unfortunately, this can also create pollution problems because HC vapors react in sunlight to form photochemical smog.

Gasoline vapors from the fuel tank and carburetor bowl, called *evaporative emissions,* can account for up to 20 percent of the emissions from an uncontrolled vehicle. To make matters worse, fuel evaporation continues around the clock regardless of whether or not the vehicle is being driven. The problem becomes especially bad on warm summer days when a car is parked in the sun on an open parking lot. Under such circumstances, gasoline vapors literally spew out of a car—unless they are prevented from doing so by an evaporative emission control system.

SEALED SYSTEM

The easiest way to control evaporative emissions is to seal off the fuel tank and carburetor bowl from the atmosphere (Fig. 4-1). But doing so is not as simple as it sounds. For one thing, a fuel tank must be vented so that air can enter to replace fuel as the fuel is used up. If a tank were sealed tight, the fuel pump would soon create enough negative suction pressure inside the tank to cause the tank to collapse or to restrict the flow of fuel to the engine. It is something like trying to pour oil out of a can with a single small hole in it. The vacuum created inside the can slows the flow of oil to a trickle. Punch a vent hole in the can and the oil gushes out. A fuel tank needs the same kind of ventilation so that the fuel pump can suck the fuel out. The job of venting the gas tank is usually performed by the gas cap.

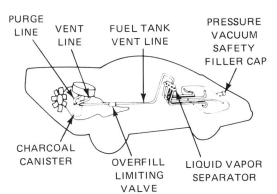

Figure 4–1 Basic components of the evaporative emissions control system.

A fuel tank must also allow for a certain amount of fuel expansion. Gasoline, like other liquids, expands as it gets warmer. If you fill a tank on a cool morning, rising temperatures later in the day can cause the fuel to expand. If there is not sufficient reserve capacity built into the tank to handle the added fuel volume, the tank will overflow.

Like the fuel tank, the carburetor bowl must also be vented to function properly. If the bowl were sealed tight, negative pressure inside the bowl would decrease the flow of fuel through the metering circuits and venturis, causing a leaning of the air/fuel mixture or possibly fuel starvation. So some means of venting the bowl must be provided.

Evaporation of fuel from the carburetor bowl increases with temperature. The hotter the bowl, the faster the rate of evaporation. While the engine is running, fresh fuel entering the bowl has a cooling effect. This helps to minimize evaporation somewhat. But as soon as the engine is turned off, the carburetor begins to soak up heat like a sponge. Evaporation increases dramatically, and on especially hot days the fuel can literally boil in the bowl.

To keep the gasoline in the fuel system and out of the atmosphere, the evaporative emission control system must allow for fuel expansion, tank venting, carburetor bowl venting, and be able to store gasoline fumes for extended periods of time.

THE NAME GAME

A variety of evaporative emission control systems have been created by the automobile manufacturers. The first systems were installed on new cars sold in California in 1970, and then on all new cars since 1971. Although these systems go by a number of different names, most share elements in common and the same basic principle of operation:

FTVC Fuel tank vapor control, AMC
ECS Evaporation control system, Chrysler
EEC Evaporative emissions control, Ford

EFE Early fuel evaporation control, GM, Datsun

EECS Evaporative emissions control system, GM

FVCR Fuel vapor control recovery, imports

Knowing the system names is not important because the manufacturers will call different systems by the same name, or the same system by different names. Terminology and consistency have always been a problem in the auto industry. For example, GM's EECS system is not the same for all models. A Chevette's system may have a somewhat different configuration than that on a Cadillac. Each auto manufacturer usually has a number of variations on the basic theme of evaporative emissions control to suit the needs of their different engine/fuel system/body styles within their product line.

The best method to determine the system specifics for your particular automobile is to refer to a service manual. The official factory shop service manuals are the best guides because they contain more detailed information than the large generalized repair manuals such as those of Motor, Chilton, and Mitchell. The big manuals cover five to seven years and every make of car, so the information must be condensed as much as possible. This sometimes means leaving out less essential service details.

Let's now look at some of the components of a typical evaporative emission control system.

FUEL TANK

All fuel tanks in today's cars are designed to allow for fuel expansion. The expansion space is usually 10 to 12 percent of the total tank volume. For example, a tank designed to hold 20 gallons of fuel when filled would need an additional 2-gallon capacity for expansion.

There are several ways expansion space can be designed into a fuel tank. The easiest way is to locate the filler neck so that an air space is created at the top of the tank when it is filled. Designing a bulge or dome on the top of the tank serves the same purpose. The air pocket absorbs the increase in volume as the fuel expands.

Another way to create an air space in the top of the tank is to connect a fill control tube to the filler neck (see Fig. 4–2). When the tank reaches a certain level as it is being filled, gasoline begins to flow back through the fill control tube into the filler neck. This causes the gas nozzle to kick off and prevents overfilling the fuel tank. The remaining air space at the top of the tank then serves as the expansion reserve.

Some vehicle manufacturers solve the expansion problem by using a small external expansion tank on top of the main fuel tank (Fig. 4–3). The expansion tank has a capacity of about 2 gallons, and is connected to the main tank with vent lines and a fill control tube. With this approach, the main tank can be filled to capacity. The expansion tank then handles any resulting fuel expansion. It is something akin to the expansion tank on a radiator.

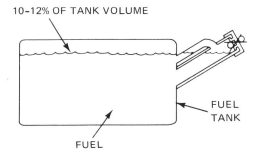

AIR SPACE PROVIDED FOR
FUEL EXPANSION

10–12% OF TANK VOLUME

FUEL
TANK

FUEL

Figure 4-2 The fill control tube between the fuel tank and filler neck prevents overfilling of the tank. An air space is needed at top to allow for fuel expansion.

EXTERNAL EXPANSION CHAMBER

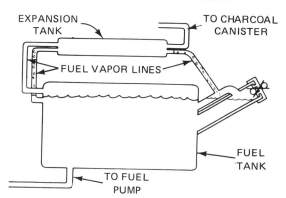

EXPANSION
TANK

TO CHARCOAL
CANISTER

FUEL VAPOR LINES

FUEL
TANK

TO FUEL
PUMP

Figure 4-3 An externally mounted expansion tank.

  Yet another approach to controlling fuel expansion is to use the tank-within-a-tank method. A little fuel tank with several small orifices punched in the sides is located inside the main fuel tank. The orifices limit the speed with which the smaller tank can fill with fuel. When the main tank is being filled, it will reach capacity long before the smaller tank. So when the gas nozzle kicks off, indicating a "full" tank, gasoline will continue to seep into the smaller tank from the big one. This creates an air space in the top of the main tank for fuel expansion.

  The only problems any of these expansion control techniques can create are complaints about slow filling. Many motorists quickly discover that such fuel tanks fill slowly or that they never seem to be quite full. That is because the tanks are designed that way. Overfilling by continually squeezing in a few more cents' worth of gasoline after the nozzle has kicked off defeats the design purpose of expansion control.

GAS CAP

Few people think of a gas cap as being an emission control device, but it is. In precontrol days, the gas cap's main job was to keep gasoline from sloshing out of the tank. It was also equipped with a small vent hole so that the tank could breathe. Air could enter through the cap to make up for fuel as it was used, and fuel vapors could exit through the cap as internal pressure built up on warm days.

Today's emission control gas caps are considerably different. They are either of solid construction (venting is provided by other means) or they contain a pressure/vacuum valve (Figs. 4–4 and 4–5). The valve-type cap will vent tank pressure if it exceeds $\frac{1}{2}$ to 1 psi. It will also allow air to enter the tank if a $\frac{1}{4}$- to $\frac{1}{2}$-inch vacuum exists within the tank. In other words, the valve-type cap can vent pressure or relieve vacuum as the situation warrants without allowing gasoline vapors to pollute the environment.

The valve itself is a simple double-spring arrangement similar to a radiator cap. One spring reacts to internal pressure while the other reacts to external pressure. A plate or diaphragm between the two springs opens and closes to allow air to pass through the valve in the direction needed.

Internal fuel tank pressure can also be vented by means of a three-way valve in the vapor line to the charcoal canister (Fig. 4–6).

PRESSURE–VACUUM RELIEF CAP
TANK PRESSURE 1/2–1 PSI

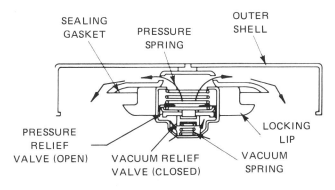

SEALING GASKET   PRESSURE SPRING   OUTER SHELL

PRESSURE RELIEF VALVE (OPEN)   VACUUM RELIEF VALVE (CLOSED)   VACUUM SPRING   LOCKING LIP

Figure 4-4 Typical emissions gas cap. Under normal conditions, the cap vents are closed preventing gasoline fumes from escaping into the atmosphere. But when internal tank pressure exceeds ½ to 1 psi due to thermal expansion, the cap vents the excess pressure to the atmosphere.

CAP RELIEVING TANK VACUUM
VACUUM 1/4″–1/2″ HG

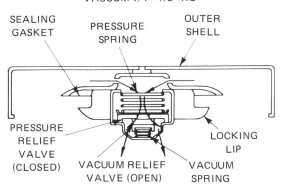

SEALING GASKET   PRESSURE SPRING   OUTER SHELL

PRESSURE RELIEF VALVE (CLOSED)   VACUUM RELIEF VALVE (OPEN)   VACUUM SPRING   LOCKING LIP

Figure 4-5 As fuel is drawn from the tank, a partial vacuum can be created. In this case, the gas cap vents pressure into the tank from outside.

OPERATION OF 3-WAY VALVE
FUEL TANK TO CANISTER

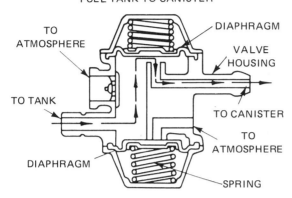

Figure 4–6 Some Ford vehicles do not have a pressure-vacuum relief gas cap. Instead they use a three-way valve in the fuel tank vent line to control internal tank pressure. The valve is shown venting tank pressure to the charcoal canister. When there's a vacuum in the tank, the upper diaphragm will allow air to be drawn into the vent line to the tank. The lower diaphragm serves as a safety vent for excessive tank pressure in case the main vent line becomes clogged.

LIQUID-VAPOR SEPARATOR

On top of the fuel tank or as part of the expansion tank is a device known as a *liquid-vapor separator* (see Fig. 4–7). The purpose of this unit is to prevent liquid gasoline from entering the vent line to the charcoal canister (located in the engine compartment). You do not want liquid gasoline going directly to the charcoal canister because it would quickly overload the canister's ability to store fuel vapors.

The liquid-vapor separator works on the principle that vapors rise and liquids sink. The vapor vent lines from the fuel tank that go to the separator are positioned vertically inside the unit with the open ends near the top. This allows the vapors to rise to the top of the separator. Any liquid that enters the separator through the vent lines (from fuel sloshing around inside the fuel tank as a result of hard driving, parking on a steep hill, excessive fuel expansion, etc.) dribbles down the sides of the vent tubes and collects in the bottom of the separator. A return line allows the liquid gasoline to dribble back into the fuel tank. The vapors then exit through an opening in

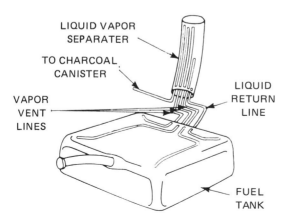

Figure 4–7 A typical liquid vapor separator.

the top of the separator, which usually has an orifice restriction to help prevent any liquid from getting into the canister vent line.

The liquid-vapor separator is a simple device that is relatively trouble-free. The only problems that can develop are if the liquid return becomes plugged with debris such as rust or scale from inside the fuel tank; if the main vent line becomes blocked or crimped; or if a vent line develops an external leak due to rust, corrosion, or metal fatigue from vibration.

Some liquid-vapor separators use a slightly different approach to keeping liquid fuel out of the canister vent line. A float and needle assembly is mounted inside the separator. If liquid enters the unit, the float rises and seats the needle valve to close the tank vent.

Another approach sometimes used is a foam-filled dome in the top of the fuel tank. Vapor will pass through the foam but liquid will cling to the foam and drip back into the tank.

If a blockage occurs in the liquid-vapor separator or in the vent line between it and the charcoal canister, the fuel tank will not be able to breathe properly. Symptoms include fuel starvation or a collapsed fuel tank on vehicles with solid-type gas caps. If you notice a whoosh of pressure in or out of the tank when the gas gap is removed, suspect poor venting.

You can check tank venting by removing the gas cap and then disconnecting the gas tank vent line from the charcoal canister. If the system is free and clear, you should be able to blow through the vent line into the fuel tank. Blowing with compressed air can sometimes free a blockage. If not, you will have to inspect the vent line and possibly remove the fuel tank to diagnose the problem.

Some Chrysler and Ford cars have an *overfill limiting valve* in the vent line between the fuel tank and charcoal canister (Fig. 4–8). The valve's purpose is to pre-

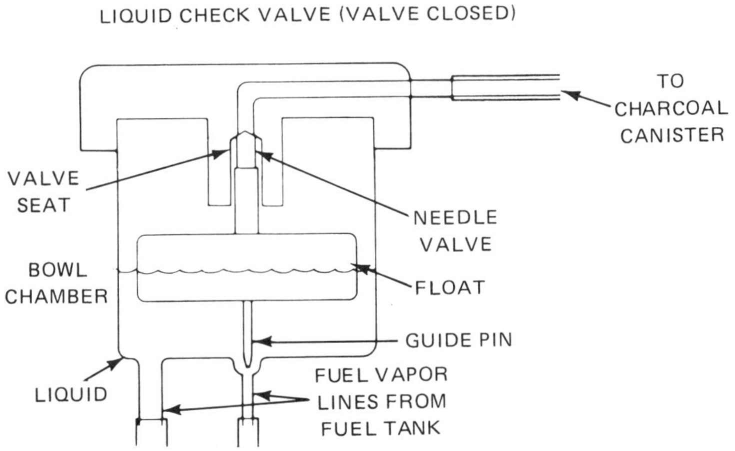

LIQUID CHECK VALVE (VALVE CLOSED)

Figure 4–8 The liquid check valve in the tank vent line closes to prevent liquid from reaching the charcoal canister.

vent any liquid that might have passed through the separator from reaching the canister. It consists of a simple float valve that closes if liquid fills the small chamber around it.

CHARCOAL CANISTER

The charcoal canister is a small round or rectangular plastic or steel container mounted somewhere in the engine compartment (Fig. 4–9). The canister's job is to store gasoline vapors from the fuel tank so that the fumes do not pollute the atmosphere. The canister contains activated charcoal—about a pound and a half of it. Charcoal acts like a sponge to soak up the gasoline vapors, holding up to twice its weight in fuel. The vapors are stored in the canister until the engine is started. The vapors are then drawn into the air cleaner, through a vacuum port in the carburetor or intake manifold, or are siphoned off through the PCV plumbing.

Some early Chrysler evaporative control systems did not use a charcoal canister. Instead, the fuel vapors were routed into the engine's crankcase for storage. When the engine was started, the PCV system would draw the fumes out of the crankcase and into the intake manifold. This approach had its drawbacks, though. For one thing, the gasoline vapors tended to dilute the crankcase oil. The vapors also formed

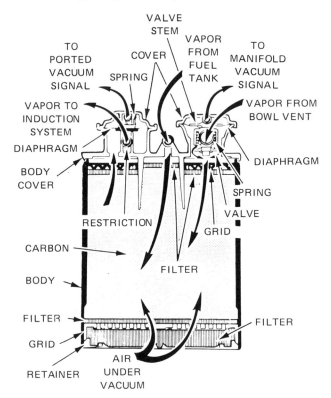

Figure 4-9 Typical charcoal canister. The charcoal soaks up and stores gasoline vapors like a sponge. The vapors are then sucked into the engine when it is started. (Ford Motor Co.)

an explosive mixture that could literally blow the valve covers right off the engine. Because of such problems, the approach was dropped in favor of the charcoal canister method of storing fuel vapors.

The charcoal canister is connected to the fuel tank via the tank vent line, and to the carburetor bowl with another vent hose. The pressure created by evaporating fuel drives the vapors through the lines and into the canister. Here they stay until the canister is purged by starting the engine.

The auto manufacturers have been quite clever in coming up with various ways to purge the charcoal canister of its contents. Some canisters have an open bottom with a small, flat air filter across the opening (Fig. 4–10). The filter is there to keep dirt out of the canister while it is being purged. The filter should be inspected periodically and replaced according to the manufacturer's recommendations. A good rule of thumb is to replace the filter every two years.

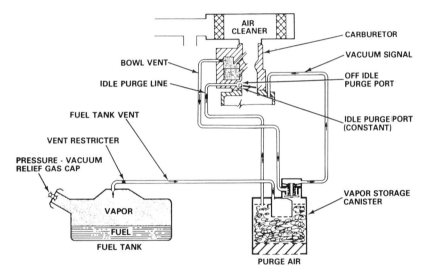

Figure 4-10 Typical charcoal canister system with an open bottomed canister. Purge air is drawn up through the canister when the purge control valve opens, allowing manifold vacuum to suck the gasoline fumes into the engine. (Chevrolet)

On canisters with the open bottom, fresh air is drawn in through the filter by connecting a purge line from the top of the canister to the air cleaner, a carburetor vacuum port, or the PCV plumbing. Airflow through the canister, *purging* as it is called, is regulated by a purge control valve on the canister. The valve opens in response to a ported vacuum signal. Others use an electric solenoid purge control valve. On these systems, the solenoid is regulated by the engine control computer.

On canisters with sealed bottoms (Fig. 4–11) fresh air is circulated through the canister via a center tube. Air is sucked down the tube, up through the charcoal, and out the purge line.

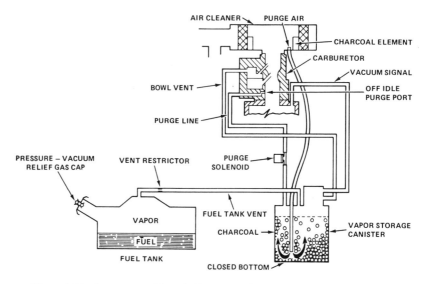

Figure 4-11 On systems with sealed bottom canisters, the air passes down through a central vent tube in the canister, and then circulates back up through the canister picking up gasoline fumes as it goes. (Chevrolet)

Depending on the design of the canister, purging may be controlled in different ways. Those that use ported vacuum use a technique called *constant and demand purge.* The purge valve on the canister allows constant purging at a restricted rate through an orifice until a certain level of vacuum exists at the canister outlet. When ported vacuum is applied to the purge control valve, it allows a higher purging rate. The reason for having constant and demand purging is because the engine cannot handle a large flow of air through the canister at idle or slow speeds. The additional air and fuel vapor would upset the air/fuel ratio, causing a rough idle and increased tailpipe emissions. At higher rpm rates, however, the engine can digest the canister's contents without problem. So purging is calibrated to match the engine's ability to handle it. A vapor feed rate of around 12 cubic feet per hour might be average for a small V8 cruising down the highway.

TROUBLESHOOTING THE CHARCOAL CANISTER

Under normal circumstances, the charcoal canister causes few problems. Since the charcoal does not wear out, it should last the life of the vehicle. Problems can result, though, when the canister filter is neglected, when a purge control valve malfunctions, or when some idiot gets the vent, purge, and control vacuum lines mixed up (the connections are usually labeled to avoid such mistakes).

If vapor is not being purged from the canister, the purge valve may be defective or the canister filter may be plugged. The purge valve can be tested with a hand vacuum pump. It should hold vacuum for at least 15 to 20 seconds without leaking down. Vacuum connections should be inspected to make sure that they are tight and properly routed.

On units equipped with computer-controlled solenoid purge valves, you will have to refer to the manufacturer's shop manual for the specs on when and under what conditions the solenoid is supposed to open. Generally speaking, such systems will not purge the canister until the engine reaches operating temperature. The coolant sensor monitors temperature and when the computer reads the appropriate temperature, it sends a command to open the canister purge solenoid.

On late-model General Motors products with Computer Command Control system, for example, the computer energizes the canister solenoid when the engine is operating in the *open-loop* mode (''open loop'' means that there is no feedback computer control over the air/fuel mixture. It is set at a fixed value until the engine warms up to improve cold idle. This prevents vacuum from reaching the purge valve on the canister. When the engine enters the *closed-loop* mode of operation (when the oxygen sensor is hot enough to produce a signal and the engine is at operating temperature), the solenoid is deenergized and vacuum is allowed to open the canister purge valve and purge the fuel vapors.

ANTI-PERCOLATOR VALVE

To prevent fuel evaporation from the fuel bowl during engine operation, some carburetors have an anti-percolator valve, where the canister vent line connects to the fuel bowl. The valve seals off the vent line while the engine is running. The valve is connected to the throttle linkage so that it will be closed when the throttle is open, and open when the throttle is closed. The valve opens when the engine is off (the throttle closed) so that the hot fuel vapors can boil out through the vent line and into the canister.

SYSTEM CHANGES

Evaporative emissions have been minimized by designing carburetors with smaller fuel bowls. The change to fuel injection in recent years has also eliminated the carburetor as a source of emissions. Fuel injection does not require venting because it has no fuel bowl. The fuel is sealed inside a pressurized system.

5

Heated-Air Intake: Blending to the Engine's Taste

The natural principle that causes a gaily-colored hot-air balloon to rise is easily grasped. The sea of air we live in is made up of a combination of gases, each of which has weight, as little as that might be. If the air is cold, the molecules of these gases are crowded together and a given volume of air will weigh more and contain more molecules than if the air is hot. That is the basic reason for the appearance of heated air intake (also known as a thermostatic air cleaner) systems on emission-controlled engines.

CARB CRACKDOWN

Early in their battle against automotive pollution, the engineers realized that the carburetor could make or break their efforts, so they clamped down on it hard. Before controls, carburetors were calibrated to be rich enough to accommodate wide variations in the temperature of the air that entered them, so they provided a mixture that would fire dependably regardless of the time of year and the vagaries of the weather. But antipollution carburetors were built to meter less gasoline into each cubic foot of air in order to reduce HC and CO emissions. The mixture became lean, and if the density of the incoming air was such that it made the blend even leaner, it would go over the edge into a ratio that simply contains too little gasoline to burn, and hesitation, stalling, and generally poor drivability resulted.

SAME VOLUME, DIFFERENT NUMBERS

If we go back to our initial illustration on the relatively small number of molecules in warm air when compared to cold air, the problem is easy to understand: Once the choke is open, a carburetor supplies the same amount of gasoline regardless of the temperature of the air that rushes through it. But the mixture—the air/fuel ratio—can still be affected by the density of the air. That is, the number of oxygen molecules (the only component of air that we are concerned with here) in a cubic foot or a gallon of cold air is greater than the number in that same volume of hot air. Therefore, introducing cold air into the engine is tantamount to introducing a larger amount of air (even though the volume is the same). This raises the ratio of air to fuel, making the blend leaner.

ENDLESS SUMMER

So what was needed with tightly controlled carburetors was a way of supplying the engine with air of nearly constant temperature (and hence, density) in both summer and winter. The lean calibration of carburetors in this era of pollution fighting simply cannot tolerate wide variations in the warmth of the intake air without leaning the blend so much that it will not burn dependably.

Enter the heated-air-intake system (Fig. 5–1). Its purpose is to duct a combina-

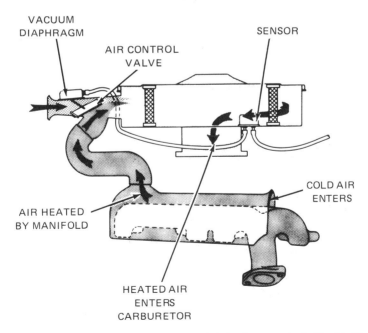

Figure 5–1   A typical vacuum-operated heated air intake system looks like this. The air is warmed as it passes over the exhaust manifold. (Chrysler)

tion of heated and unheated air to the carburetor to make sure that the power plant is provided with air of the proper density for smooth, responsive running regardless of the ambient or engine compartment temperature. It eliminates the tendency of lean mixture calibrations to cause roughness, stumbling, and missing in cool weather or before the engine has warmed up.

BEYOND THE CHOKE

Of course, the choke acts to richen the mixture for a short time after startup, but if it were designed to stay on all the time on cold days, emissions would soar while mileage and power would plummet. So the heated-air-intake system takes over where the choke leaves off. An ancillary benefit is the virtual elimination of carburetor icing.

ANATOMY

A typical heated-air-intake system (Fig. 5–1) comprises a sealed chamber with a diaphragm and spring inside (called a *vacuum motor*), linkage that connects the diaphragm to a trap door or flapper valve in the air cleaner snorkel, a metal shroud or stove mounted over the exhaust manifold, a heat tube that runs from the shroud to the bottom of the air cleaner snorkel where the flapper valve is, a thermostatic vacuum bleed valve inside the air cleaner (Fig. 5–2), a vacuum line that connects the bleed valve to the vacuum motor, and another vacuum line that goes to the intake manifold.

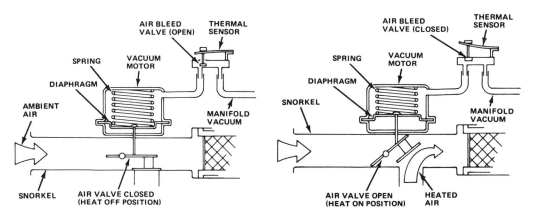

Figure 5–2   A thermostatically controlled bleed valve controls the vacuum flow to the vacuum motor. When the temperature is low, the bleed valve is closed and full vacuum reaches the vacuum motor which raises the flapper valve allowing only heated air to enter the air cleaner. As the temperature rises, the bleed valve opens gradually, reducing the strength of the signal to the vacuum motor. (American Motors Corp.)

OPERATION

The operation of this setup is easy to understand. When the engine is not running, no vacuum is present and the spring inside the vacuum motor holds the linkage down. This, in turn, keeps the flapper valve in the down position where it closes off the passage that connects to the heat tube and manifold shroud.

When the engine is started, it naturally generates vacuum in its intake manifold. This is routed to the thermostatic bleed valve. This valve is closed below a specified temperature, and in this condition it allows vacuum to be applied to the vacuum motor.

This raises the diaphragm in the motor against spring pressure and lifts the flapper valve so that it closes off the snorkel to unheated air and opens the passage from the heat tube and shroud. In this position, the only air the engine gets to run on must pass over the exhaust manifold, which gets hot very quickly, and so is warmed above ambient temperature.

As the engine warms up, the bimetallic strip inside the thermostatic bleed valve starts to respond. It moves the valve off its seat gradually, opening the circuit between the intake manifold and the vacuum motor. In other words, it allows vacuum to bleed off, lessening the strength of the signal to the vacuum motor.

This reduces the pull of the diaphragm inside the vacuum motor against the spring, and the spring forces the linkage downward opening the snorkel to unheated air.

PROGRESSIVE

The action here is not simply on or off, but a progression that blends heated and unheated air. Once a certain temperature is reached, only unheated air enters the engine.

When the accelerator pedal is floored and the throttle plates are opened all the way, vacuum in the intake manifold drops. So even if the engine is cold, too little vacuum acts on the diaphragm to overcome the spring and the flapper valve falls, admitting only dense, unheated air into the engine to blend with the rich mixture present during acceleration and maximum potential power is produced. Variations on this theme include a separate cold-air passage into the air cleaner that is kept closed by vacuum.

Another type of full-throttle control is exemplified by Ford's Cold Weather Modulator (Fig. 5-3). This is a thermostatically controlled check valve that traps vacuum in the vacuum motor when the car is accelerated hard at temperatures below 55 °F. This eliminates hesitation and flat spots by allowing heated air to enter the engine in spite of the drop in vacuum that naturally occurs when the throttle is opened wide. It is the opposite of most systems that allow only unheated air to flow to the carburetor when the accelerator is floored.

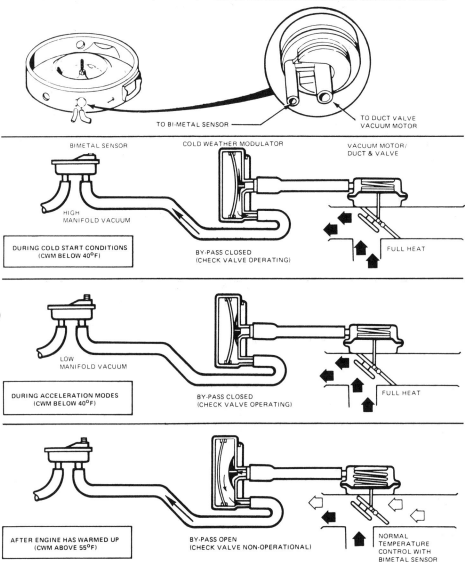

PREVENTS VACUUM SIGNAL OVERRIDE AND CONTINUES TO PROVIDE HEATED AIR TO THE CARBURETOR UNTIL THE COLD WEATHER MODULATOR REACHES 55°F, AFTER WHICH THE BIMETAL SENSOR CONTROLS DUCT AND VALVE DOOR FUNCTION.

TO BI-METAL SENSOR

TO DUCT VALVE
VACUUM MOTOR

BIMETAL SENSOR   COLD WEATHER MODULATOR   VACUUM MOTOR/
DUCT & VALVE

HIGH
MANIFOLD VACUUM

DURING COLD START CONDITIONS
(CWM BELOW 40°F)

BY-PASS CLOSED
(CHECK VALVE OPERATING)

FULL HEAT

LOW
MANIFOLD VACUUM

DURING ACCELERATION MODES
(CWM BELOW 40°F)

BY-PASS CLOSED
(CHECK VALVE OPERATING)

FULL HEAT

AFTER ENGINE HAS WARMED UP
(CWM ABOVE 55°F)

BY-PASS OPEN
(CHECK VALVE NON-OPERATIONAL)

NORMAL
TEMPERATURE
CONTROL WITH
BIMETAL SENSOR

Figure 5-3 The Ford Cold Weather Modulator valve is spliced into the hose between the bleed valve and the vacuum motor. It traps vacuum at temperatures below 55° F. so that, even during wide open throttle operation when manifold vacuum is low, the flapper valve stays in the heat-on position and the engine is fed only heated air. (Ford)

Some cars do not use vacuum to control air blending. Instead, they employ a direct-acting thermostatic arrangement. The American Motors version has a thermostatic bulb that is connected to the rod that operates the flapper valve in the air cleaner snorkel (Fig. 5–4). As the thermostat warms up, it pushes on the rod, closing the door to the heated air from the exhaust manifold shroud and allowing unheated air to enter the air cleaner. The operation of this setup cannot be as precisely controlled as that of a vacuum-activated system, but it is simpler.

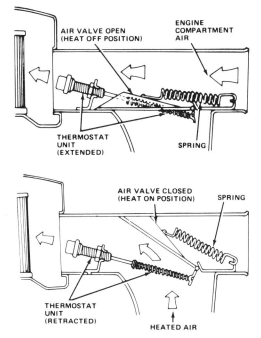

Figure 5-4 Some vehicles use direct-acting thermostatic units that open and close the flapper valve without the help of vacuum. (American Motors Corp.)

ILLS

The symptoms associated with a malfunction in the heated-air-intake system are exactly what you would expect: hesitation, stalling, and rough idling if too little hot air is ducted to the carburetor, and conversely, reduced power output and fuel efficiency if the flapper valve keeps the unheated air passage closed.

Since a leak or other interruption in the vacuum signal to the vacuum motor, or a perforated diaphragm in the vacuum motor itself will normally result in the flapper valve closing off heated air (in other words, always open to unheated air), the first set of problems just mentioned is the most common. Only if there is some sort of binding condition or if the bleed valve jams closed will the flapper valve stay up and close off unheated air.

EXAM

Testing a typical system is simple. First, give it a visual examination. Make sure that the vacuum lines are not cracked, crushed, or broken, and that they are properly connected. Next, see that the heat tube that runs from the exhaust manifold stove to the air cleaner snorkel is intact. Look inside the snorkel—you may have to remove the air cleaner or use a mirror—to see if the flapper valve is in the down position. Alternatively, you can insert a pencil or screwdriver to ascertain the position of the valve.

Start the engine. Provided that the temperature is below that at which the bleed valve opens, the flapper valve should rise to the heat-on position, then drop gradually as the engine warms up.

If you do not get these results, there are several ways to proceed to uncover the fault. One logical progression is as follows:

1. Attach a spare line to a known source of manifold vacuum or to a hand-operated vacuum pump (Fig. 5–5).
2. Remove the standard vacuum line from the nipple of the vacuum motor.
3. Connect your source of vacuum to the vacuum motor.
4. If the flapper valve rises now, the trouble is probably in the bleed valve or the line that feeds it vacuum.
5. If the flapper valve does not rise when given direct vacuum, there is most likely a leak in the vacuum motor diaphragm or possibly the flapper valve or linkage is jammed.

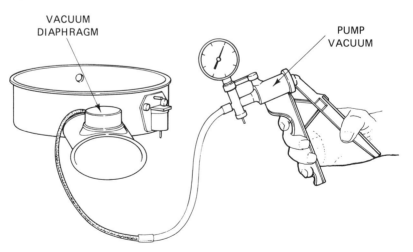

VACUUM
DIAPHRAGM

PUMP
VACUUM

Figure 5–5 The vacuum motor diaphragm and the flapper valve can be tested with a hand-operated vacuum pump. When vacuum is applied to the motor, the valve should rise. (Chrysler)

6. If you are using a hand-operated vacuum pump and the flapper valve does rise when vacuum is applied to the vacuum motor, see that the motor holds vacuum for a reasonable length of time (a common limit is less than a 10-in. Hg drop in 5 minutes). If not, the diaphragm is leaking.

7. To check the bleed valve, first look up the temperatures at which it should close and open. A common Chrysler system, for example, will raise the flapper valve with a cold engine and the ambient temperature below 50 °F.

8. Remove the air cleaner and cool the bleed valve below the closing temperature. This can be done by covering it with ice cubes.

9. With the hand pump, apply vacuum to the inlet side of the bleed valve (make sure that the outlet side is properly connected to the vacuum motor).

10. If the flapper valve does not rise, the bleed valve is probably defective. It is relatively inexpensive and very easy to replace.

NONVACUUM TYPE

The normal procedure for checking the nonvacuum type of thermostatically controlled air cleaner is, first, to see that the flapper valve and its linkage move without binding, then to immerse the thermostatic bulb in cold water to see that it raises the flapper valve, and hot water to see that it lowers it.

The specific steps for a typical example, the AMC six-cylinder, are as follows:

1. Remove the top of the air cleaner and immerse the entire snorkel in a large pan of cold water, making certain that the thermostatic bulb is covered.

2. Place a thermometer in the pan and observe it while heating the water slowly.

3. The flapper valve should remain closed below 105 °F. At 130 °F, the valve should be open completely.

4. If this does not occur, and the valve mechanism is not binding, the thermostatic bulb is defective.

6

Air Injection:

Force Feeding O₂

Very soon after the industry-wide campaign against air pollution from motor vehicles began, it became apparent that designing a gasoline piston engine that would burn up every trace of HC and CO inside its cylinders was analogous to achieving the scientists' elusive absolute zero: Progress could be made toward the goal, but complete success was impossible.

NATURALLY DIRTY

No matter how cleverly the combustion chambers were shaped to eliminate quench areas, or how precisely controlled the fuel mixture calibration or the spark advance curve was, or how drastic changes in valve timing, temperature range, and bore-to-stroke ratio were, the engine would still pump a considerable amount of pollutants into our fragile atmosphere. It is simply in the nature of the familiar, dependable Otto cycle piston engine, mainly because some fuel vapor will always condense back into a liquid on the cylinder walls and the surfaces of the combustion chamber, and only the vapor will burn. This results in the emission of a certain amount of HC, which is basically unburned gasoline, and CO, which is a product of incomplete combustion.

Even very efficient designs that allowed an overall lean mixture, such as Honda's CVCC (Compound Vortex Controlled Combustion) system, could not circumvent the laws of physics, although they went a long way. Clearly, there was, and is, no chance that any gasoline-burning engine as we know it could emit only CO_2 and

water vapor from its exhaust ports, the by-products of pe█
dentally, the same ingredients that, with sunlight, make p█

CRUTCHES

So the engineers had to create some crutches to hold up this basic inadequ█
otherwise very practical engine design. These would have to be add-on syste█
treated the exhaust after it left the engine.

The first of these was quite successful: The air injection system (Fig. 6–1), whi█
appeared on many cars in 1968 and was soon dubbed the "smog pump" by those
who like to come up with amusing nicknames. It cleaned up the pollutants in the
exhaust quite thoroughly and cost very little in terms of efficiency. With the advent
of catalytic converters, many cars use this system to provide the extra air the catalyst
needs to work properly.

A typical version of the system (Fig. 6–2) comprises a belt-driven vane-type air
pump (Figs. 6–3 to 6–5), the plumbing necessary to route the pressurized air into

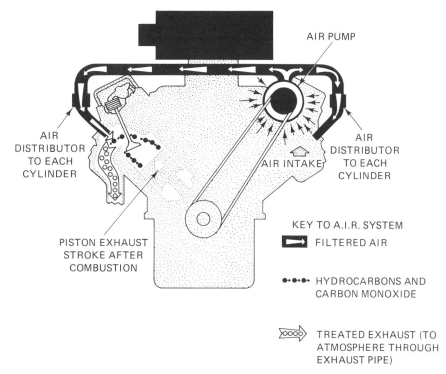

AIR PUMP

AIR DISTRIBUTOR TO EACH CYLINDER

AIR INTAKE

AIR DISTRIBUTOR TO EACH CYLINDER

PISTON EXHAUST STROKE AFTER COMBUSTION

KEY TO A.I.R. SYSTEM

▬► FILTERED AIR

•+•+•► HYDROCARBONS AND CARBON MONOXIDE

⟫⟩ TREATED EXHAUST (TO ATMOSPHERE THROUGH EXHAUST PIPE)

Figure 6–1 This schematic of a basic air injection system shows how the pump directs the fresh air into the exhaust ports. (Buick)

fect combustion (and inci-
ants grow).

acy of an
ns that

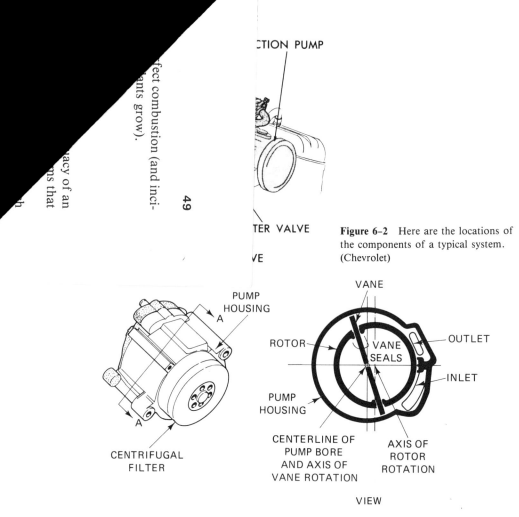

...CTION PUMP

...TER VALVE

...VE

Figure 6-2 Here are the locations of
the components of a typical system.
(Chevrolet)

PUMP
HOUSING

A

CENTRIFUGAL
FILTER

VANE

ROTOR

VANE
SEALS

PUMP
HOUSING

CENTERLINE OF
PUMP BORE
AND AXIS OF
VANE ROTATION

OUTLET

INLET

AXIS OF
ROTOR
ROTATION

VIEW

Figure 6-3 The pump is of the sliding-vane variety and is driven by a belt.
(Chevrolet)

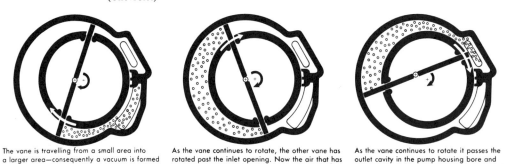

The vane is travelling from a small area into
a larger area—consequently a vacuum is formed
that draws fresh air into the pump.

As the vane continues to rotate, the other vane has
rotated past the inlet opening. Now the air that has
just been drawn in is entrapped between the vanes.
This entrapped air is then transferred into a
smaller area and thus compressed.

As the vane continues to rotate it passes the
outlet cavity in the pump housing bore and
exhausts the compressed air into the remainder
of the system.

Figure 6-4 A sliding-vane pump moves air as shown here. (Chevrolet)

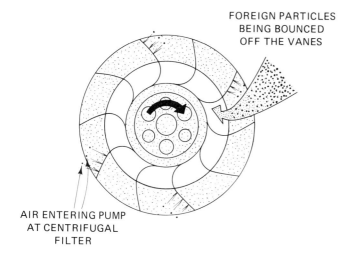

FOREIGN PARTICLES
BEING BOUNCED
OFF THE VANES

AIR ENTERING PUMP
AT CENTRIFUGAL
FILTER

Figure 6-5 A centrifugal filter behind
the pulley keeps dirt out of the pump.
(Chevrolet)

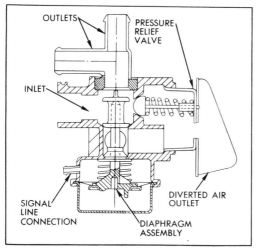

OUTLETS
PRESSURE
RELIEF
VALVE
INLET
SIGNAL
LINE
CONNECTION
DIVERTED AIR
OUTLET
DIAPHRAGM
ASSEMBLY

Figure 6-6 A typical diverter valve
directs pump output to the exhaust
ports or the atmosphere according to a
vacuum signal from the intake manifold. (Chevrolet)

the exhaust ports, a check valve to keep exhaust gases from backing up into the pump,
and a diverter or bypass valve (Fig. 6-6) that can shunt the flow to the atmosphere
under certain conditions of engine vacuum.

MODUS OPERANDI

The operation of an air injection system is as follows. When the engine is started,
a belt causes the pump shaft pulley to rotate. Inside the pump, vanes riding against
the walls of a cylindrical chamber start moving air (Fig. 6-4). The rotor turns on
an axis that is different from that of the pump bore, and the vanes slide in and out
of slits in the rotor. This causes volume variations that result in the pumping action.
Usually, intake is through a centrifugal filter mounted behind the drive pulley (Fig.

6-5), but separate intake filters have been used on some applications. Most systems use a relief valve that allows excess pump pressure to escape. It may be in the pump itself, or incorporated into the diverter valve (Figs. 6-6 and 6-7). The pressurized air exits into a large-diameter hose that routes it to the diverter valve. This valve can switch airflow from the hose that connects to the check valve and air manifold to a small muffler that exits into the atmosphere. It has a vacuum chamber, diaphragm, and spring setup that moves the valve from one of its seats to the other and so controls the switching operation. The vacuum chamber is connected to the intake manifold or the base of the carburetor by a hose. During deceleration when vacuum in the intake manifold is very high, the diaphragm in the valve's chamber overcomes the force of the spring and switches airflow from the outlet to the air manifold to the muffler, then dumps it into the atmosphere (Fig. 6-8). This prevents the backfiring in the exhaust system that would occur if the extra air the pump provides were available to allow the rich mixture that is present in the exhaust stream during deceleration to ignite explosively. To put it simply, the diverter valve shunts the air pump's output away from the exhaust system during deceleration.

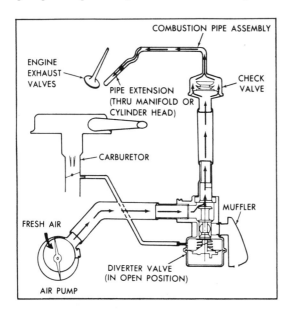

Figure 6-7 During most modes of operation, the diverter valve routes pump output to the exhaust stream. (Chevrolet)

During all modes of engine operation except deceleration, the diverter valve allows the air from the pump to flow into the hose to the check valve. This is a simple one-way device, which lets air enter the air injection manifold, but keeps exhaust pressure from backing up into the diverter valve and the pump if a belt should break or the pump should otherwise stop working (normally the pump's pressure is high enough to overcome exhaust pressure). From the check valve, the air flows into the air injection manifold, which directs it into each exhaust port. Here, as we said above, it gives the HC and CO in the exhaust oxygen to combine with.

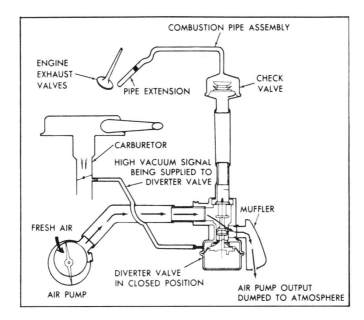

Figure 6-8 During deceleration, vacuum is high, and the strong signal to the diverter valve causes it to dump pump output through the muffler into the atmosphere. This prevents the explosive ignition of the rich decel mixture in the exhaust system commonly known as backfiring. (Chevrolet)

ALTERNATIVES

Other means of preventing backfiring are sometimes used instead of the diverter valve. Many late models have both a vacuum differential valve (VDV) and an air bypass valve working together (Fig. 6-9). The VDV shuts off the vacuum signal to the bypass valve momentarily whenever intake manifold vacuum rises or drops sharply. In normal operation while the bypass valve is receiving vacuum, it allows air pump output to flow freely to the air injection manifold. When the VDV stops the vacuum signal, a spring inside the bypass valve opens a vent port and pump flow is shunted to the atmosphere. If there is excessive pump pressure or a restriction in the check valve (Fig. 6-10) or air injection manifold, a relief valve inside the bypass valve opens, allowing some of the pump's output to be dumped and the rest of it to take the normal route.

GULP VALVE

Another antibackfire device is known as the gulp valve. It has a diaphragm chamber, a spring, and a normally closed valve inside. In a typical specimen, during deceleration when intake manifold vacuum reaches 20 to 22 in. Hg, the vacuum signal pulls

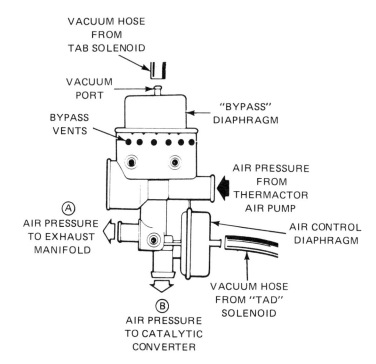

VACUUM HOSE
FROM
TAB SOLENOID

VACUUM
PORT

BYPASS
VENTS

"BYPASS"
DIAPHRAGM

AIR PRESSURE
FROM
THERMACTOR
AIR PUMP

Ⓐ
AIR PRESSURE
TO EXHAUST
MANIFOLD

AIR CONTROL
DIAPHRAGM

Ⓑ
AIR PRESSURE
TO CATALYTIC
CONVERTER

VACUUM HOSE
FROM "TAD"
SOLENOID

Figure 6-9   This combination air bypass and air control valve eliminates backfiring and switches air pump output between the exhaust ports and the three-way catalytic converter. The bypass function is basically the same as that of a diverter valve except that it shunts air to the atmosphere when its vacuum signal is cut off. (Ford)

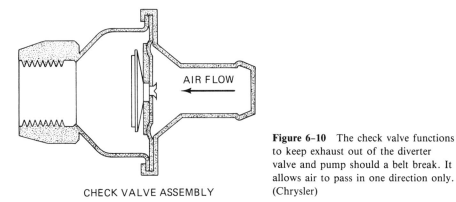

AIR FLOW

CHECK VALVE ASSEMBLY

Figure 6-10   The check valve functions to keep exhaust out of the diverter valve and pump should a belt break. It allows air to pass in one direction only. (Chrysler)

the diaphragm against spring pressure opening the valve. This allows some of the air pump's output to flow into the intake manifold to dilute the rich mixture present in this mode and thereby eliminates the possibility of extra gasoline vapor exploding in the exhaust system. Usually, the addition of pump air to the intake stream lasts from 1 to 3 seconds.

THERMAL REACTOR

Before the catalytic converter came into general use, it looked like another device might be the answer to the emissions problem: the thermal reactor (Fig. 6–11). Basically just a larger and heavier exhaust manifold, it was supposed to give the HC and CO in the exhaust a much greater opportunity to combine with oxygen supplied by a high-capacity air injection setup. It had the advantage over the catalyst of not requiring the use of expensive metals (chiefly platinum and palladium), and unleaded gasoline. But it did not do the job as well as the catalytic converter did, and it raised underhood temperatures considerably, so the catalyst won out. It was, however, adopted for a few cars, notably the rotary engine-powered Mazda RX–7, which used it until 1981, when a switch to a catalytic converter was made. We are including a mention of the thermal reactor here because it illustrates the furthest the air injection principle can be taken.

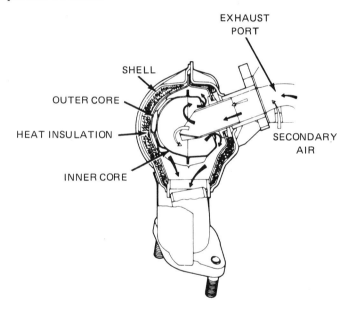

Figure 6-11   This Dodge Colt thermal reactor shows the extent to which the principle of burning up pollutants at the exhaust ports can be taken. (Dodge)

FEEDING THE CAT

In 1975 when most cars sold in this country were manufactured with catalytic converters, the purpose of the air injection system changed somewhat. Now it is used primarily to give the pollutants in the exhaust stream sufficient fresh air to combine with inside the converter. The mixture of HC, CO, and O_2 from the air pump flares

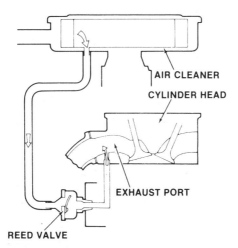

Figure 6-12 Aspirator valve systems make use of the vacuum between pulsations in the exhaust system to inject fresh air. (Dodge)

up in the presence of the catalyst. Temperatures as high as 1600°F are generated, eliminating most of the harmful pollutants by oxidizing them into CO_2 and water.

Some catalyst-equipped cars have been built that did not need air injection. The engines are calibrated in such a way that there is enough O_2 in the exhaust to support the reaction in the converter. But these are exceptions. Most vehicles today have either an air injection system or an aspirator-valve setup (Fig. 6-12) to provide extra fresh air to let the catalyst do its job.

AIRFLOW SWITCHING

Some cars employ a system that switches air injection flow from the exhaust ports to a line that runs either directly into the catalytic converter itself or into the pipe ahead of it under certain conditions.

The 1980 Ford Electronic Engine Controls (EEC) III system is an example of this (Fig. 6-13). In it, the Electronic Control Assembly (ECA, a microprocessor computer) makes decisions on where to send Thermactor (air pump) output on the basis of input from various engine sensors, notably that for coolant temperature, and according to certain time calibrations. It works in conjunction with a pair of solenoid valves: the bypass solenoid, which shunts air pump flow to the atmosphere when energized, and the diverter solenoid, which switches airflow to either the exhaust ports or the catalytic converter.

During normal engine operation, air pump output is routed to the catalytic converter (there is a mixing chamber inside the shell between the reduction catalyst and the conventional two-way oxidation catalyst, and this is where the air enters). The computer energizes the bypass solenoid when time at closed throttle exceeds a specified time in its memory, when the interval between a lean and a rich signal from the oxygen sensor is longer than a set time value, and during wide-open throttle. This

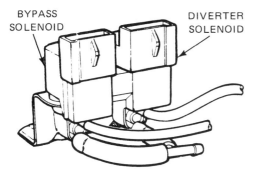

BYPASS
SOLENOID

DIVERTER
SOLENOID

Figure 6-13 Through the use of solenoid operated vacuum valves, electronic signals from a computer can control air injection. (Ford)

protects the catalyst from the danger of overheating from the presence of an overly rich mixture, and guards against backfiring.

The computer energizes the diverter solenoid when the coolant sensor tells it that the engine is cold. This directs air pump flow upstream to the exhaust manifold during engine warmup, giving HC and CO more time to oxidize.

A MECHANICAL VARIATION

Air switching is accomplished on some Chrysler cars without the help of electronics. An ordinary diverter valve is used to dump air pump output into the atmosphere during deceleration. But the shunting of flow from the exhaust ports downstream to the exhaust pipe ahead of the catalytic converter is accomplished by an air switching valve and a coolant control engine vacuum switch (CCEVS). The purpose of this setup is different from that of the Ford system: When the engine is cold, fresh air directed at the exhaust ports is very helpful in reducing HC and CO emissions. But after normal operating temperature is reached, the high heat levels generated here start producing NO_x. If the air is injected downstream in the exhaust system, where it is cooler, HC and CO will still be oxidized, but NO_x emission will not be adversely affected.

The switching action is accomplished as follows. As long as there is a vacuum signal from the CCEVS to the air switching valve, the valve stays open, allowing air pump output to flow to the exhaust ports. When the engine warms up, the CCEVS shuts off the vacuum signal and the switching valve shunts most of the airflow to the downstream injection point. A bleed hole in the valve allows a small amount of air to be routed to the exhaust ports at all times, which assists in the reduction of HC and CO, but is not enough to promote the manufacture of NO_x.

AIR SANS PUMP

A phenomenon little known outside engineering circles made a simplified variation on the air injection theme possible. It seems that a partial vacuum is created in the exhaust manifold momentarily after each cylinder's exhaust stroke. This is the result of the exhaust valve's closing and the inertia of the column of spent gases as it speeds through the exhaust system. After the valve has closed, the column continues to travel away from the exhaust port, leaving a negative pressure area behind it. This will be easier to understand if you picture the exhaust not as a steady stream, but as a series of pulses.

This principle has been used on some cars to eliminate the air injection pump and its small, but constant parasitic loss. It also cost less money, reduced complication, and saved space in that crowded engine compartment.

Basically, the system consists of a one-way valve (or valves) that is connected to the exhaust manifold in such a way that exhaust cannot flow out, but the air can flow in (Figs. 6-14 and 6-15). The generic name for this device is *aspirator valve*. When the partial vacuum mentioned above occurs in the exhaust manifold, atmospheric pressure opens the valve and fresh air is admitted to combine with the hot gases. Whenever a pressure pulse appears in the exhaust system, it forces the valve closed so that there are no noisy and objectionable leaks. It is much more effective at idle and low-speed operation than at high rpm levels.

GM's Pulsair is a common example of this system. It comprises a housing that contains one valve for each cylinder (Figs. 6-15), a tube that goes into each exhaust manifold runner, and a hose that picks up fresh air from the air cleaner.

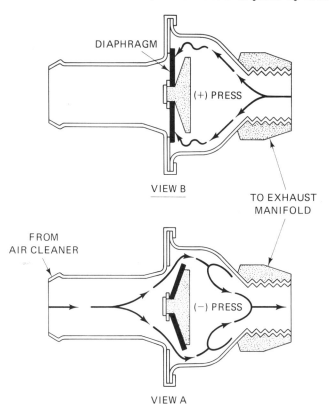

Figure 6-14 An aspirator is simply a one-way valve that allows air into the exhaust manifold, but won't let exhaust gases out. (Chrysler)

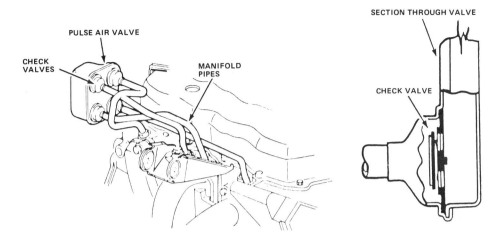

Figure 6–15 This Pulsair system uses one aspirator valve for each cylinder. (American Motors Corp.)

7

Catalytic Converters

The hurried, almost frantic efforts of automakers to eliminate the air pollution their vehicles caused paid off handsomely. The redesigns and controls worked, so much so that by the early 1970s, a typical domestic car was emitting 85 percent less HC, 70 percent less CO, and 50 percent less NO_x than a comparable conveyance from the pre-emission-control era, albeit with a 15 percent or more loss of fuel efficiency.

NOT GOOD ENOUGH

But that was not enough to satisfy the government. The Clean Air Act of 1970 demanded more. The goal it stipulated seemed impossible to reach, and the auto manufacturers seemed to be fighting a losing battle. If they could not meet the standards, they would not be allowed to sell their products in the United States, and our economy's basic industry would be crippled. The ramifications would be too terrible to contemplate. The car companies' executives cried: "You can't legislate science and engineering!" Did you ever try to get the very last drop out of a bottle? You can get most of it, even almost all of it, but every last drop? The situation looked hopeless.

But Detroit had a card up its sleeve that it did not play until things got critical: the amazing catalytic converter. With some cooperation from the oil companies and the government, this gave the automakers a winning hand, or at least allowed them to stay in the game.

SECRET WEAPON

Air injection was the first add-on crutch used to treat exhaust gases after they left the cylinders, and it helped a great deal (in fact, it is still retained in many configurations). But the catalytic converter was a much more powerful weapon—an add-on, a crutch to be sure, but one so effective that it allowed other emission control systems to be recalibrated so that fuel efficiency and drivability could be improved.

A catalyst is a strange thing. It is defined as any substance that increases the rate of a chemical reaction without being changed one whit in the process. It stands by and encourages chemical activity without being used up.

In the case of the ordinary two-way automotive catalyst (Figs. 7-1 and 7-2), the activity is the oxidation of HC and CO, which it assists immensely without being altered itself. It consists of a superthin coating of platinum and palladium on pellets or a honeycomb of ceramic material and a durable stainless steel housing that is spliced into the exhaust system between the manifold and the muffler.

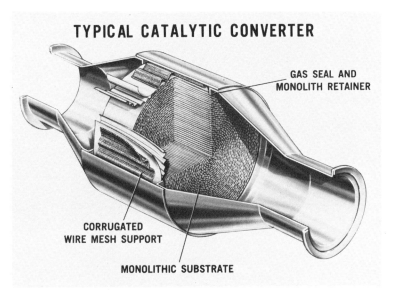

TYPICAL CATALYTIC CONVERTER

GAS SEAL AND
MONOLITH RETAINER

CORRUGATED
WIRE MESH SUPPORT

MONOLITHIC SUBSTRATE

Figure 7-1 Here's a typical monolithic catalytic converter. The honeycomb substrate is ceramic with a "noble metal" coating. (Champion)

The hot exhaust gases pass over these "noble" metals on the substrate and miraculously flare up and combine with the extra oxygen that is available because of an air injection system or a very lean mixture or both, and ordinary, harmless CO_2 and water (the same things that, with sunlight, make plants grow) are released into the atmosphere (Fig. 7-3). Barely a trace of HC and CO are left by the time the exhaust reaches the tailpipe.

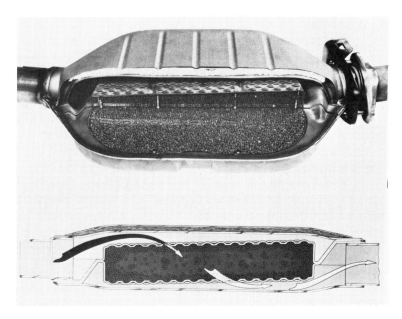

Figure 7-2   This is a pellet bed-type catalytic converter. The ceramic pellets are coated with the catalytic material. (AC)

CATALYTIC CONVERTER

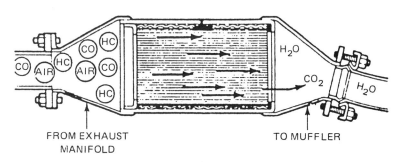

Figure 7-3   With an oxidation catalyst, HC, CO, and O_2 enter and are converted to H_2O and CO_2, two harmless substances. (Ford)

This type of catalytic converter—called the *oxidation catalyst*—appeared in 1975 on about 90% of the automobiles built in America. Since that time, it has become standard equipment on almost every new car sold in this country whether domestic or foreign (except, of course, for diesels).

TRIPLE CAT

In the last few years, a more complex variation on this theme has showed up: the three-way reduction type of catalytic converter system. This does the same thing as the two-way oxidation type, but it adds another catalytic material—rhodium—that splits troublesome NO_x into innocuous nitrogen and oxygen. It requires that the air/fuel ratio be much more precisely regulated, however, so it is usually used only on cars with closed-loop/electronic feedback carburetion or fuel injection (that is, a system that employs an oxygen sensor and microprocessing to alter the mixture as necessary to keep it as nearly perfect as possible).

OXIDATION TYPE

There are two basic types of two-way oxidation catalytic converters: honeycomb (Fig. 7–1) and pellet (Fig. 7–2). The former consists of a ceramic core that is coated with a microscopically thin layer of the catalytic agent. The configuration of the core gives it immense surface area so that the pollutants are bound to come into contact with the platinum.

The latter is filled with ceramic pellets. They have a thin platinum coating and a large composite surface area, too. The difference is that if they become contaminated with lead or otherwise rendered inoperative, they can be dumped and replaced by a new load. The honeycomb type would have to be junked altogether.

Both kinds of converters have a stainless steel skin, so they will not rust out. Since the temperature inside gets up to 1600 °F in normal operation, most installations include heat shields of one kind or another (Fig. 7–4) to avoid the possibility of setting the carpet on fire or melting the driveway.

Some cars have minioxidation or quick-light-off honeycomb catalysts mounted in the exhaust pipe ahead of the main converter (Fig. 7–5). Basically, they are just smaller versions of a regular cat, and their purpose is to get the reaction started sooner.

SUPPORTING THE BLAZE

Plenty of oxygen is needed to keep the fire going in the converter, so catalyst-equipped cars have air injection, a very lean blend, or an exhaust pulse type of air inlet system to make sure the cat gets enough of that combustion supporting gas.

The main problems these great launderers have are fouling, as when regular gas is put into the tank (Fig. 7–6) and the lead additive covers up the platinum, and overheating, (Fig. 7–7) which usually results from a misfiring plug or an overly rich mixture that supplies the catalyst with so much stuff to oxidize that the reaction gets out of hand. The rotten-egg smell? Well, unless it is constant, you will just have to learn to live with it. It is supposed to decrease after the cat has been seasoned for a few thousand miles.

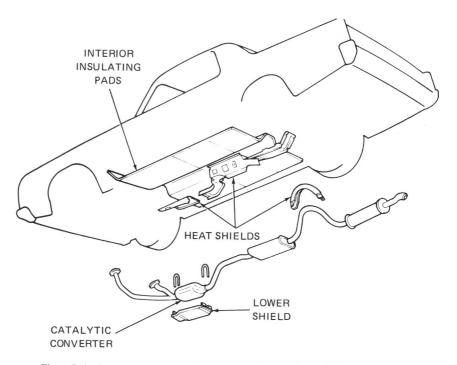

Figure 7-4 Converters can operate at up to 1,600° F., so heat shields are necessary. (Chrysler)

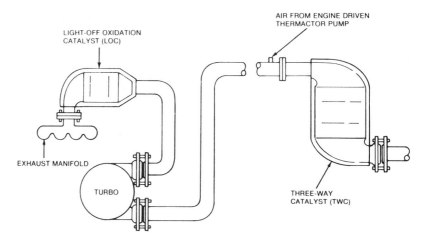

Figure 7-5 On some applications, like this turbocharged Ford, a small light-off catalyst is added to handle emissions right after the engine is started. It gets hot enough very quickly. (Ford)

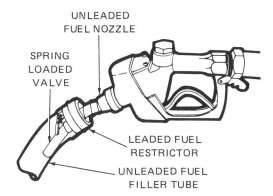

UNLEADED
FUEL NOZZLE

SPRING
LOADED
VALVE

LEADED FUEL
RESTRICTOR

UNLEADED FUEL
FILLER TUBE

Figure 7-6 Since the thin coating of catalytic material would be rendered ineffective by a coating of lead, only unleaded gasoline may be used in cat-equipped cars. To avoid mistakes, fuel fillers were made too small to accommodate regular pump nozzles. (Chrysler)

Figure 7-7 This core overheated and melted due to an excessively rich mixture. The backpressure was so great that the car had hardly a shadow of its former power. (Champion)

REDUCTION CATALYSIS

Of the terrible trio—HC, CO, and NO_x—the last has been the most troublesome automotive pollutant to eliminate. One of the ingredients of irritating and unwholesome photochemical smog, NO_x is short for oxides of nitrogen, a group of chemicals that contain both nitrogen and varying amounts of oxygen. NO_x is contrary. The things the engine designers did to reduce the production of HC and CO actually promoted the formation of NO_x. Extra heat and oxygen helped burn up the other two, but those conditions are exactly what is needed to create NO_x.

Then we got EGR and different camshaft grinds. These combined to dilute the incoming charge so that its peak burning temperatures were not as high. Electronic

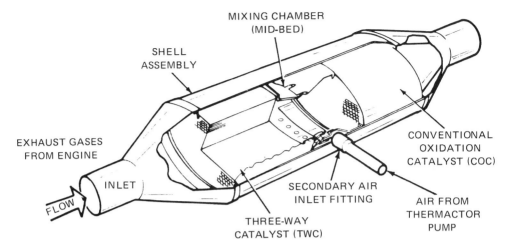

Figure 7–8   Dual or three-way catalytic converters handle NO_x along with HC and CO. The front portion is a reduction catalyst which uses rhodium to break NO_x into N and O_2, and the rear portion is a conventional oxidation catalyst. Air is injected into the mixing chamber downstream of the reduction cat so that it doesn't upset the exacting combination of gases that's needed for NO_x control. (Ford)

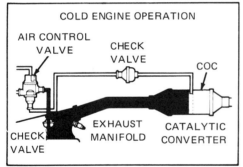
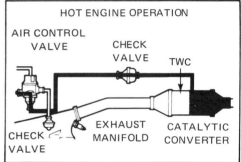

Figure 7–9   There is very little NO_x production when the engine's cold, so the air control valve switches air injection flow to the exhaust manifold to help oxidize the high levels of HC and CO that are present during cold running. After the engine has warmed up, air flow is shunted to a point downstream of the reduction portion of the catalyst so that NO_x control can begin. (Ford)

spark advance control helped too, and NO_x finally started to retreat. But only so far. And Big Brother said, "Not far enough."

So the automakers adopted the three-way catalyst (Figs. 7–8 and 7–9). It allowed them to keep building cars without having to add so much EGR that fuel efficiency was murdered.

DELICATE

Unfortunately, the reduction reaction is very delicate. To keep it going at an acceptable level, the air/fuel ratio fed into the engine has to be nearly perfect (Fig. 7–10).
That means a ratio of 14.6:1 air to fuel by weight (regardless of how rich or lean
the blend is that goes into the engine, the actual burning of gasoline always occurs
at this ratio), a mixture called *stoichiometric*.

   If an engine were only asked to run at a constant speed, temperature, and load,
a simple carburetor could be calibrated to supply this theoretically ideal blend. But
cars operate in the real world, and that means the power plant has to contend with
a constantly changing set of demands and circumstances. The finest carburetor can
not come anywhere close to maintaining a stoichiometric mixture. The best fuel injection systems can get within $\pm$ 3 percent, but even that is not close enough for
good reduction.

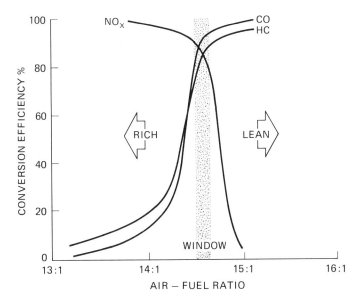

Figure 7–10   The efficiency of the three-way catalyst is dependent upon the air/fuel
ratio. If the ratio is not held very closely to 14.6:1, the catalyst's ability to cut pollutants
falls dramatically. (Volkswagen)

CLOSED-LOOP

What was needed was a setup that would ensure that the mixture stayed stoichiometric
no matter what the conditions, something that continually altered the amount of fuel
metered into the intake stream so that it always stayed in the same proportion to
the volume of air.

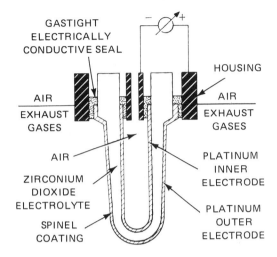

Figure 7-11 In order to maintain the proper air/fuel ratio, closed loop or feedback systems were designed, and they depend upon voltage signals from an oxygen sensor. (Volkswagen)

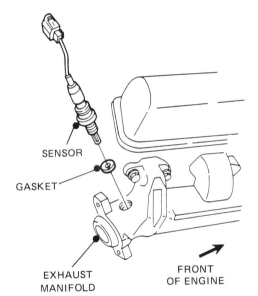

Figure 7-12 The exhaust gas oxygen sensor is screwed into the exhaust manifold. (Ford)

And that is where the closed-loop idea comes in (Fig. 7-11). By using an oxygen sensor (Figs. 7-12 and 7-13), the proper electronic logic, and a feedback carburetor or EFI, the fuel system could be made to monitor itself, continually making adjustments to keep the mix precise.

Some cars, the Mazda GLC, for instance, manage to get decent reduction from the three-way cat without resorting to closed-loop, but they are exceptions. For really efficient catalysis, it is absolutely necessary.

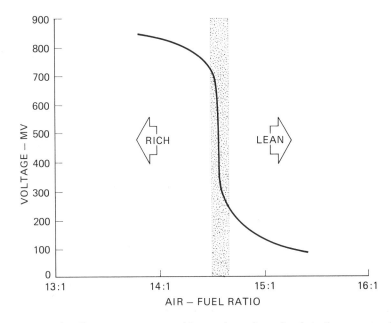

Figure 7-13   The oxygen sensor provides varying voltage signals to the computer according to the oxygen content of the exhaust. The computer then commands the carburetor or fuel injectors to correct the mixture. (Volkswagen)

SWITCHING

After all the effort and expense of keeping the blend so close to perfect, it would be idiotic to pump air into the exhaust manifold and throw the ratio out of bounds before it reached the reduction catalyst. So a typical three-way converter has the reduction portion upstream of the oxidation section with a mixing chamber between (Fig. 7-11). The extra air the conventional catalyst needs to do its job is routed from the pump into the mixing chamber, so it does not affect the reaction that splits NO_x.

Since NO_x formation is not a problem until the engine warms up, there is usually a themostatically controlled valve that switches air pump flow from the catalyst to the exhaust manifold under cold-running conditions, and this helps burn up the huge amounts of HC and CO the engine produces during this mode (Fig. 7-12). After normal operating temperature is reached, flow is routed to the catalyst.

So what does the three-way catalytic converter do for us when it is all boiled down? It allows us to drive around in cars that are more fuel-efficient than anyone would have believed possible a decade ago, while emitting next to no pollutants. It is an important advance, and there is no indication that we will ever go back to the way things were in the simple (if wasteful and dirty) old days.

8

Exhaust Gas Recirculation

Like the PCV system and the catalytic converter, exhaust gas recirculation (EGR) is one of the basic emissions control systems with which every mechanic should be familiar.

CONTROLLING NO$_x$

Nearly 80 percent of the air we breathe is nitrogen. By itself, nitrogen is a harmless gas that normally does not do much of anything. But at temperatures above 2500 °F, such as those found inside a combustion chamber, the nitrogen reacts with oxygen to form a variety of toxic compounds. One such compound is nitrogen dioxide, a brownish gas that is one of the main ingredients in smog. The haze that obscures Los Angeles on a sunny day is a prime example of NO$_x$ pollution. It irritates the eyes and lungs, and makes breathing difficult. It is 15 times as deadly as carbon monoxide, and in concentrations of as little as 100 to 300 parts per million it can cause a person to cease breathing altogether. In lesser dosages, say 25 to 75 ppm, it can cause the lungs to fill with fluid. Prolonged exposure to even low concentrations of NO$_x$ can cause a general deterioration in health, including loss of weight and poor digestion. It has even been known to bring on pneumonia in some people.

Convinced that NO$_x$ is nasty stuff? So was the Environmental Protection Agency when it decided to establish NO$_x$ emissions standards for automobiles back in the early 1970s. At the time, engineers were designing engines with later ignition timing, leaner carburetion, and so on, to raise combustion temperatures so that less HC and CO would be produced in the exhaust. Unfortunately, they discovered that the things which reduced HC and CO emissions (namely, higher combustion temperatures) created more NO$_x$. Figuring out how to reduce HC and CO while reducing NO$_x$ seemed like an impossible task—until engineers found that a long-duration hot burn with lower peak or flash temperatures would keep HC and CO down while minimizing NO$_x$ formation. The solution was to dilute the air/fuel mixture with a small amount of noncombustible gas. The diluted mixture burned slower and produced lower peak temperatures, exactly what was needed to reduce all three pollutants.

The next problem was to figure out how to accomplish the same results on mass-produced engines. A system was needed to inject a small part of noncombustible gas into the intake manifold. That was the easy part. The hard part was coming up with an endless supply of inert gas. The answer turned out to be exhaust gas recirculation. The engine manufactures plenty of inert noncombustible gas in the form of exhaust. Since exhaust is mostly carbon dioxide and water vapor, there is practically no oxygen or gasoline left to burn again.

To reroute some of this exhaust gas back into the intake manifold, passages are cast into the intake manifold and plumbing is added to the exhaust manifold. The heart of the system is a gadget called an EGR valve. Its job is to meter the amount of exhaust gas that enters the intake manifold according to demand.

The first EGR system appeared on certain 1972 Buicks, and then on practically all cars since 1973. Although other methods of controlling NO$_x$ have been tried, such as increasing camshaft valve overlap, redesigning combustion chambers, and modifying ignition advance curves, these methods by themselves are generally not adequate to meet the strict emissions standards. They do, however, add to the effectiveness of the EGR system. Only a few vehicles (Saab 99E is one) have been able to meet NO$_x$ standards without resorting to EGR.

NO$_x$ or KNOCKS?

EGR has another little recognized affect on engine performance, and that is to prevent detonation. The "cooling effect" that EGR has on combustion temperatures means that the engine can handle more spark advance without detonation (pinging or spark knock caused by too much spark advance and/or low-octane fuel). The more spark advance an engine can have, the more power and economy it will deliver.

If the EGR system is rendered inoperative, as when someone decides to unhook all the "pollution junk," the cooling effect is lost. Now the engine will have too much spark advance for the stock timing setting. The result is spark knock during acceleration or when the engine is heavily loaded. Severe detonation can lead to serious problems, such as broken rings, cracked pistons, and rod bearing damage. If ignition timing is retarded relative to stock specifications, performance and fuel economy will suffer. The only other alternative is to switch to high-octane gasoline (which is more costly and which may not entirely eliminate the problem). The best answer, of course, is to make sure that the EGR system is functioning properly and not to tamper with it (which is illegal for professional mechanics).

EGR VALVE: THE SOLUTION TO NO$_x$ POLLUTION

The *EGR valve* is what makes the EGR system work (Fig. 8–1). The unit consists of a vacuum diaphragm and a little plunger valve. The end of the plunger valve, which may be either a poppet or taper stem design, separates a port in the intake manifold from the exhaust system.

Under part-throttle operation in a warm engine, the valve opens and allows engine vacuum to draw a small amount of exhaust into the intake manifold (Fig. 8–2). The amount of exhaust that enters the intake manifold is determined by the

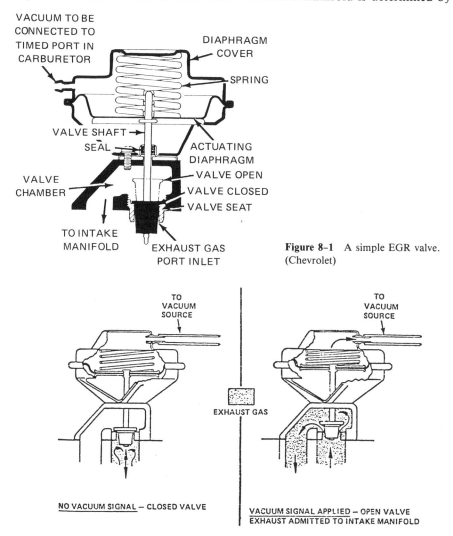

Figure 8–1 A simple EGR valve. (Chevrolet)

NO VACUUM SIGNAL – CLOSED VALVE

VACUUM SIGNAL APPLIED – OPEN VALVE
EXHAUST ADMITTED TO INTAKE MANIFOLD

Figure 8–2 Under part-throttle operation, intake vacuum is used to pull the EGR valve open so exhaust can flow into the intake manifold. (Chevrolet)

size of the EGR valve orifice through which the exhaust flows and how far the valve is held open. At maximum flow, the amount of dilution is about 6 to 10 percent exhaust in the intake manifold.

Since the exhaust contains very little oxygen and unburned fuel, mixing it with the incoming air/fuel mixture is something like mixing ashes with paper to be burned. Because there is not as much air and fuel to burn, combustion temperatures are reduced and NO_x formation is minimized—as long as the temperatures remain below 2500 °F.

PART-TIME EGR

EGR is not a full-time thing. Because it has the same effect on drivability as a small vacuum leak, it is used only during part-throttle operation. EGR is not necessary at idle because the volume of NO_x formed is fairly small. EGR at idle would also cause the engine to run rough and possibly stall (this is one symptom of an EGR valve that is stuck in the open position). EGR is not wanted during engine cranking because again it acts like a vacuum leak and would cause hard starting. Neither is EGR used during full-throttle or wide-open-throttle operation because of its negative effect on performance.

How does the EGR valve know when to open and when to close? It operates in response to engine vacuum either from a ported vacuum source above the carburetor throttle plates (Fig. 8–3) or from venturi vacuum at the carburetor throat (Fig. 8–4).

With ported vacuum systems (Figs. 8–3 and 8–5), engine vacuum pulls the EGR valve open when the throttle plates are cracked open far enough to expose the vacuum port to intake vacuum. The use of ported vacuum prevents EGR at idle. A spring

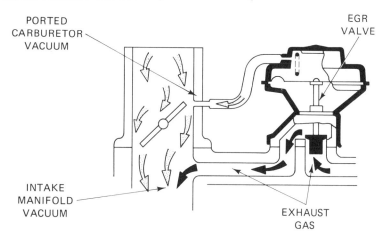

Figure 8–3 Ported vacuum is used to prevent EGR operation at idle. Only when the throttle plates are opened past the vacuum port does intake vacuum reach the valve. (Chevrolet)

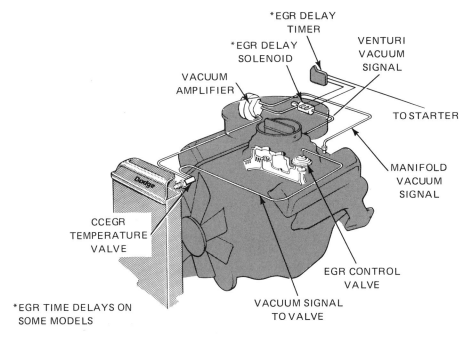

Figure 8-4 In venturi type control systems, a vacuum amplifier is needed to boost the relatively weak venturi vacuum signal. This Chrysler system shows some of the typical EGR components. It also has an EGR delay timer and delay solenoid to prevent EGR operation during starting. (Chrysler)

over the diaphragm inside the EGR valve pushes the valve shut when vacuum is cut off (as when the throttle plates close) or when engine vacuum drops below a certain level (as during wide-open-throttle operation).

EGR systems using venturi vacuum require a *vacuum amplifier* (Fig. 8-4) to boost the relatively weak vacuum signal at the carburetor venturi. Reading vacuum from this point allows precise measurements between EGR action and airflow through the carburetor. The amplifier is connected to a vacuum reservoir in many applications, and contains a check valve to maintain an adequate vacuum supply regardless of variations in engine vacuum. A relief valve may also be used to dump or cancel the output EGR signal whenever the venturi vacuum signal is equal to or greater than intake vacuum. This allows the EGR valve to close at wide-open throttle when maximum power is required.

Another "trick" that has been incorporated into EGR systems is the ability to change EGR flow in response to changes in exhaust system back pressure. NO_x formation increases when the engine is under load and during acceleration, so EGR flow should be increased to compensate for the higher operating temperatures. Since exhaust back pressure is a good measure of engine load, having a pressure-sensitive diaphragm that reacts to changes in back pressure is an effective means of regulating EGR operation. The back-pressure diaphragm, which may be located inside the EGR

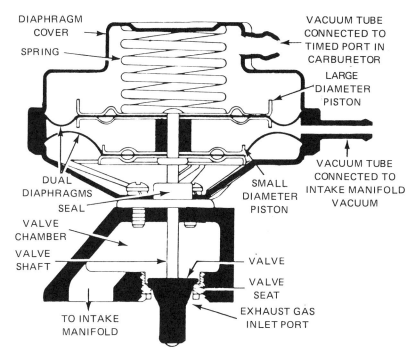

Figure 8-5   This is a dual diaphragm, dual vacuum hose connection EGR valve. Ported vacuum is balanced against intake vacuum to regulate EGR operation. (Chevrolet)

valve itself together with the main control diaphragm (Figs. 8–6 and 8–7), or in a separate housing (Fig. 8–8), opens and closes a small vacuum bleed hole in the main EGR vacuum circuit or diaphragm chamber. Opening the bleed hole obviously reduces the vacuum to the EGR valve and prevents it from opening fully. Closing the bleed hole allows full vacuum and maximum EGR flow. The back-pressure diaphragm, therefore, increases or decreases EGR vacuum in direct proportion to changes in exhaust back pressure.

POSITIVE AND NEGATIVE BACK-PRESSURE EGR VALVES

There are two types of back-pressure EGR valves: positive and negative. The positive type (Fig. 8–6) uses positive exhaust back pressure to regulate EGR flow. As pressure increases in the exhaust, the valve begins to open, allowing increased EGR into the intake manifold. This reduces back pressure somewhat, allowing the back-pressure diaphragm to bleed off some control vacuum. The EGR valve begins to close and exhaust pressure rises again. The EGR valve oscillates open and closed with changing exhaust pressure to maintain a sort of balanced flow. The negative back-pressure type of EGR valve (Fig. 8–7) reacts in the same way, except that it

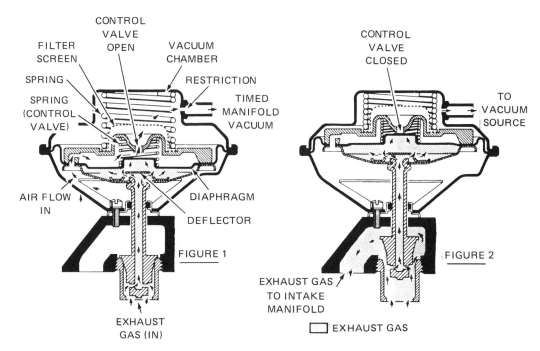

Figure 8-6 A positive backpressure EGR valve. Exhaust pressure passing up the hollow valve stem pushes against a backpressure diaphragm. As the diaphragm lifts, it closes the air bleed allowing full vacuum in the main diaphragm chamber. This causes the EGR valve to open. When pressure falls, the lower backpressure diaphragm drops, opening the air bleed in the control valve and reducing vacuum in the main diaphragm chamber. The valve then closes. Oscillating up and down, the valve keeps pace with changes in exhaust pressure. (Chevrolet)

reacts to *negative* or decreasing pressure changes in the exhaust system to regulate EGR action. Decreasing back pressure signals decreased engine load, and the back-pressure diaphragm opens a bleed hole to reduce EGR flow. It is the same principle as with the positive type except that the control function reacts to decreasing pressure rather than increasing pressure.

Back-pressure EGR valves, both positive and negative, provide better control over EGR flow because they have the ability to react directly to changing engine loads. There are a couple of drawbacks, however. For one, the small bleed hole can become clogged. On back-pressure EGR valves with the back-pressure diaphragm inside the valve itself, a hollow valve stem is used to carry exhaust pressure to the diaphragm. This stem can also become plugged quite easily. Another problem is that back-pressure-sensitive EGR valves are affected by problems or changes in the exhaust system, such as a restricted or plugged muffler or converter. Replacing the original equipment muffler or converter with a unit that does not produce the same system back pressure can also affect EGR operation.

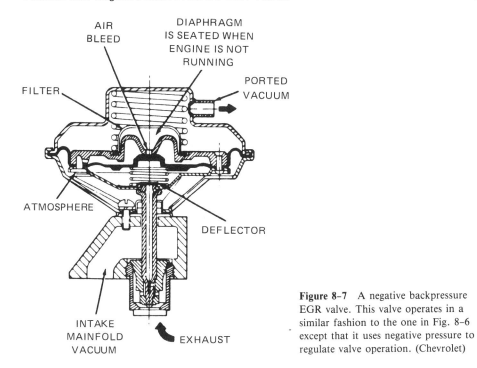

Figure 8-7 A negative backpressure EGR valve. This valve operates in a similar fashion to the one in Fig. 8-6 except that it uses negative pressure to regulate valve operation. (Chevrolet)

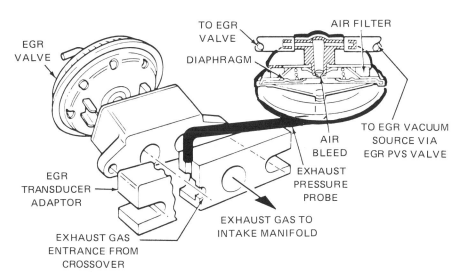

Figure 8-8 Some EGR systems have separate backpressure diaphragms to alter the vacuum signal strength to the EGR valve. The one shown here is typical of Ford. (Ford Motor Co.)

TVS SWITCH

Most pre-computer EGR systems have a *temperature vacuum switch* (TVS) or ported vacuum switch (Fig. 8–4) between the EGR valve and vacuum source to prevent EGR operation until the engine has had a chance to warm up. The engine must be relatively warm before it can handle EGR. If an engine runs rough or stumbles when cold, it may indicate a defective TVS that is allowing EGR too soon after startup. A TVS stuck in the closed position would block vacuum to the EGR and prevent any EGR operation. The symptom here would be excessive NO_x emissions and possible pinging or detonation.

WIDE-OPEN-THROTTLE SWITCH OR VALVE

Many EGR systems use some type of wide-open throttle switch or valve to cut out or eliminate EGR action during those times when maximum power and acceleration is needed. On some systems, a diaphragm vents EGR vacuum to the atmosphere when intake manifold vacuum drops to zero (indicating a wide-open throttle). On computer systems, a wide-open-throttle switch tells the computer when to cut out the EGR valve.

OTHER MEANS OF MODULATING EGR ACTION

Some systems use an air bleed orifice or solenoid to modulate EGR action. According to engine operating conditions, the air bleed may be opened to reduce EGR vacuum, which in turn reduces how far the EGR valve opens (Fig. 8–9). On some General Motors applications, for example, the amount of EGR is reduced when the automatic transmission torque converter clutch (TCC) is engaged (Fig. 8–10). The

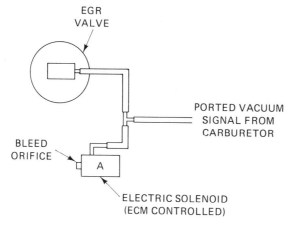

Figure 8–9   GM EGR system with one control solenoid and a bleed orifice. (Chevrolet)

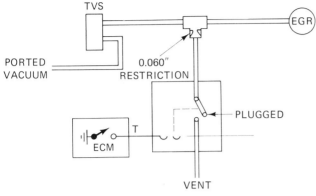

Figure 8-10 On some systems, such as the 1982 TCC EGR bleed system shown here, EGR vacuum is reduced when the transmission torque converter locks up. (Chevrolet)

same thing may be used on vehicles with manual transmissions when running in high gear. The reason for doing this is to provide smoother engine operation. Too much EGR flow can cause roughness and hesitation.

EGR COMPUTER CONTROLS

On most engines with computerized engine control, a temperature vacuum switch is not used because EGR vacuum is controlled by the computer. The computer monitors engine temperature through the coolant sensor, and when the programmed operating temperature is reached the computer opens the EGR vacuum solenoid, allowing intake manifold vacuum to pass through to the valve. On some systems, the control solenoid is normally closed. Energizing it opens the solenoid and allows vacuum to reach the EGR valve. On other systems, the EGR solenoid may be open in the normal position. It is energized (closed) only when EGR is not wanted, as when the engine is cold, during cranking, or at wide-open throttle. In either case, once the engine warms up, EGR flow is controlled as usual by ported vacuum and exhaust back pressure.

Some computer systems use air bleeds or vents in conjunction with the EGR solenoid to regulate EGR flow (Fig. 8-11). Both Ford and General Motors do this on certain systems. Ford Electronic Engine Control (EEC-II and III) systems have two EGR control solenoids (Fig. 8-12) and an EGR valve position sensor mounted on top of the EGR valve (Fig. 8-13). The EEC computer monitors various engine functions as well as EGR position to determine how much EGR is needed. Opening the normally closed EGR vacuum control solenoid (EGRC) allows manifold vacuum to pass to the EGR valve. The normally open EGR vent solenoid (EGRV) vents some air into the vacuum line to reduce vacuum and limit how far the EGR valve opens. If more EGR is needed, the computer energizes (closes) the EGRV to stop the air

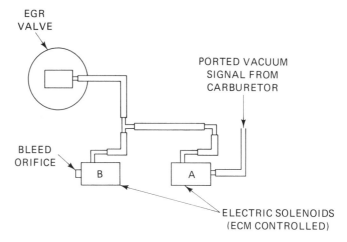

Figure 8-11   An EGR system with two computer-controlled solenoids. EGR operation is regulated by energizing one or both solenoids. (Chevrolet)

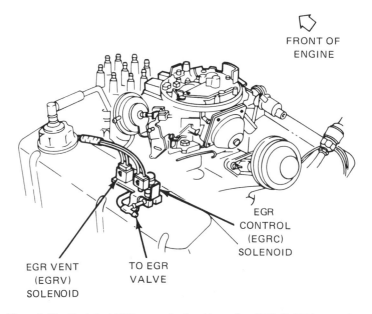

Figure 8-12   Ford dual EGR control solenoids used on EEC-II, EEC-III and some EEC-IV applications. (Ford Motor Co.)

leak so that full vacuum can reach the EGR valve. The computer can modulate EGR action by "dithering" (opening and closing) the two solenoids to achieve the amount of EGR needed. Energizing both provides maximum EGR (full vacuum), energizing only the EGRC but not the EGRV provides a sort of "midrange" EGR (part vacuum), while not energizing either solenoid prevents any EGR (no vacuum).

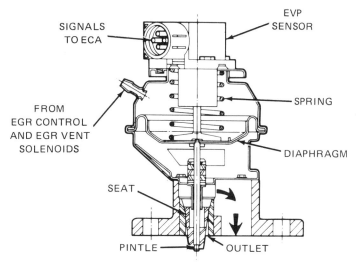

Figure 8-13 EGR valve position sensor on Ford EEC system. (Ford Motor Co.)

The EEC-II and EEC-III EGR systems are somewhat complicated, and Ford has simplified them somewhat on 1.6- and 2.3-liter EEC-IV systems by eliminating the dithering EGRV vent solenoid (both solenoids are still used on V6 and V8 engines). The 1.6- and 2.3-liter EEC-IV engines use only the single EGRC vacuum shutoff for EGR control.

Another variation on the theme is to use positive air pressure rather than intake vacuum to open the EGR valve. General Motors calls its version of this approach the Aspirated EGR system. Ford has a somewhat similar system on 1978–1979 Lincoln Versailles (EEC-I). Air pressure from the engine's air pump is diverted through the air bypass valve to the EGR valve. EGR operation is regulated by computer control of the air bypass valve and ported carburetor vacuum.

A new approach to EGR control, introduced by General Motors on certain 1984 Buick and Chevrolet Full Function Computer Command Control systems, is a *pulse-width-modulated EGR control solenoid*. By using the vehicle computer to cycle the EGR vacuum solenoid rapidly on and off, a variable vacuum signal can be produced to regulate EGR operation closely. It is the same principle that is used to change the air/fuel ratio by changing the duty cycle or dwell time of the mixture control solenoid in a feedback carburetor. The amount of "on" time versus "off" time for the EGR solenoid ranges from 0 to 100 percent, and the average amount of "on" time versus "off" time at any given instant determines how much EGR flow occurs.

With diesel engines, there is no intake vacuum for EGR operation or control, so vacuum is usually created by an auxiliary vacuum pump (Fig. 8–14). The pump provides a steady amount of vacuum for opening the EGR valve, and operation is regulated by computer-controlled vacuum solenoids and input from an electrical back-pressure sensor in the exhaust system.

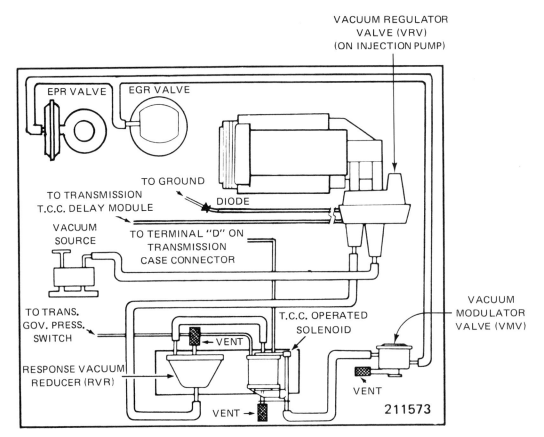

Figure 8-14   The EGR system on a GM V6 diesel. Vacuum is provided by an auxiliary vacuum pump. A vacuum regulator and exhaust backpressure sensor control EGR operation. (Chevrolet)

EGR DIAGNOSIS

With so many different variations in EGR systems, the first step in EGR diagnosis is to identify the type of system on which you are working. How is EGR operation controlled? Is it ported vacuum, venturi vacuum with a vacuum amplifier, or what? Does the system have a temperature vacuum switch, a computer-controlled EGR vacuum or vent solenoid, a wide-open-throttle switch or valve? Is the EGR valve a back-pressure-sensitive type, and if so, what type? Is there an EGR valve position sensor? Are there other systems, such as canister purge, plumbed into the EGR vacuum circuit? You can figure out these things by (1) visually inspecting the system, (2) referring to the vacuum hose routing decal under the hood, and/or (3) looking up the system description in your shop manual.

If you understand the fundamentals of how an EGR system is supposed to work (which is what we have covered in this chapter), you should be able to determine whether or not the system is functioning properly, and if not, which component is at fault. Most factory shop manuals include detailed step-by-step diagnostic procedures for troubleshooting the different systems. Computer trouble codes by themselves only tell you that you have a problem, not the exact component that may be causing it. For that, you need to follow the diagnostic procedure outlined by the vehicle manufacturer.

Keep in mind the conditions under which EGR is and is not used. It is not used during cranking, during idle, when the engine is cold, or at wide-open throttle. It is used in varying degrees under part-throttle operation.

EGR problems fall into one of several basic categories:

1. *Carbon buildup around the EGR valve seat.* This eventually prevents the valve from closing completely (allowing EGR at idle and causing rough idle and hard starting) or causes the valve to stick shut (no EGR flow, resulting in excessive NO_x emissions and spark knock). Carbon can also plug up the hollow stem and/or bleed hole on back-pressure-type EGR valves, which interferes with proper action of the valve. Carbon-fouled valves can sometimes be cleaned with a wire brush and solvent, but solvent can damage the rubber diaphragm inside the valve. If the clogging is internal, the valve must be replaced.

2. *Diaphragm failure of the EGR valve or back-pressure unit.* Rubber diaphragms deteriorate with age and with exposure to heat and corrosive chemicals such as carburetor cleaner, degreasers, and solvents. Once a diaphragm loses its ability to hold a vacuum (splits, cracks, pinholes, tears, etc.), EGR action will be adversely affected. Since nobody makes a serviceable EGR valve, the only repair you can make is to replace the EGR valve (or back-pressure control if a separate unit).

3. *Loss of vacuum signal.* Vacuum leaks or obstructions in the vacuum plumbing, a TVS, vacuum amplifier, vacuum solenoid, vacuum bleed, back-pressure diaphragm, or related vacuum system will obviously interfere with proper operation of the EGR valve.

4. *Failure of a computer-controlled solenoid.* If a solenoid fails to operate when energized, jams shut or open, or fails to function because of a corroded electrical connection, loose wire, bad ground, or other electrical problem, EGR operation will be affected. Depending on the nature of the problem, there may be no EGR, EGR all the time, or insufficient EGR. If bypassing the suspicious solenoid with a section of vacuum tubing causes the EGR valve to operate, find out why the solenoid is not working before automatically replacing it. The solenoid may be okay but an electrical problem may be at fault.

    In rare instances, a computer malfunction will prevent the EGR system from functioning properly. Internal computer problems are sometimes self-diagnosed and trigger trouble codes. In any event, the vehicle manufacturer's exact diagnostic procedure should be followed if a computer problem is suspected.

5. *Failure of a computer input sensor such as a back-pressure sensor, wide-open-throttle switch, or EGR valve position sensor.* Computer trouble codes will usually turn up for most sensor problems (Ford, for example, will show a trouble code 31 and/or 32 if there is an EGR problem) but a trouble code does not necessarily always mean that the problem is in the sensor itself. It could be in the wiring. Again, a careful diagnostic examination following the vehicle manufacturer's step-by-step instructions must be followed if an accurate diagnosis is to be made. Occasionally, a glitch in the system will interfere with EGR operation without triggering a trouble code. An EGR valve position sensor that does not change its output value much might be such an example. Ford says that voltage output should increase as the valve opens. This can be tested by reading output voltage with a digital 10-megohm impedance voltmeter while applying vacuum to the EGR valve. Voltage will increase from a few tenths of a volt up to 5 + volts if the EGR valve position sensor is working properly.

Most manufacturers recommend checking the EGR system every 24 months or 25,000 miles, or when replacing the spark plugs, to see that the system is functioning properly. The EGR system, as are all other emission control components, is covered by the five-year/50,000-mile warranty. If a failure occurs during this time, the vehicle manufacturer is obligated to make the necessary repairs free of charge, assuming that the failure was not the result of tampering, negligence, or accident damage.

EGR QUICK CHECK

To check EGR valve operation, do the following:

1. With the engine at operating temperature, rev it up to about 2000 rpm. You should be able to see the EGR valve stem move upward (a mirror and light will probably be necessary). Be careful not to touch the valve, because it will be hot.

   If the valve fails to move or oscillate, check to see that vacuum is reaching the valve when the engine is run at part-throttle. Do this by connecting a tee and vacuum gauge to the main EGR vacuum connection at the EGR valve. If vacuum is reaching the valve but it is not moving, you must assume that the valve is stuck (carbon deposits on valve stem or seat) or that the diaphragm is defective. Carbon accumulations can be cleaned, but a defective valve must be replaced.

   A word of caution about replacing EGR valves: With so many variations within any given vehicle manufacturer's product lines, it is extremely important that you install the exact replacement for the particular vehicle application on which you are working. Two EGR valves may look identical from the outside but carry different calibrations inside. Always check the part numbers carefully—and do not always assume that the old EGR valve that came off the engine is the right one for the application. Someone else may have replaced it and installed the wrong one for the application.

2. With the engine at idle, apply vacuum to the EGR valve (use either a hand-held vacuum pump or tap into the intake manifold at a convenient connection). This should cause the EGR valve to open (Fig. 8–15). Idle speed should drop at least 100 rpm or become rough if the valve is opening. If the valve does not move, remove it and apply vacuum with a hand pump to see if it moves (this works only on single-diaphragm-type valves, not those with built-in back-pressure diaphragms).

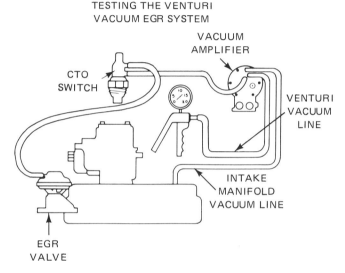

Figure 8–15 A hand vacuum pump and vacuum gauge are your best friends when it comes to troubleshooting vacuum controls on an EGR system. For computers, you'll need a volt-ohm meter to check out the control solenoid(s) and wiring connections.

To check EGR valves with external back-pressure diaphragms:

1. With the engine at operating temperature, slowly open and close the throttle, being careful not to exceed one-half throttle or 3000 rpm. The EGR valve stem should move up or down if the valve is working.

2. If the EGR valve stem fails to move, disconnect the vacuum hose from the back-pressure sensor and connect a vacuum gauge to the temperature vacuum switch. The TVS resembles a ported vacuum switch and is usually mounted somewhere on the engine block so that its tip will be in contact with the coolant inside the block. Slowly open and close the throttle while watching the vacuum gauge. If the vacuum is less than about 4 inches, the TVS is defective. If vacuum is getting through okay, tee the vacuum gauge between the EGR valve and the

back-pressure sensor. Repeat the throttle test again, slowly opening and closing the throttle while keeping the engine under 3000 rpm. If no vacuum is noted, replace the back-pressure sensor. If vacuum is getting through and readings are normal but the EGR valve stem does not move, replace the EGR valve.

To test EGR valves with internal back-pressure control:

1. With the engine at operating temperature, open the throttle gradually until the engine reaches 1500 to 2000 rpm. The EGR valve should move smoothly toward the open position, then begin to slowly cycle (stem move up and down slightly in a rhythmic pattern).
2. Release the throttle suddenly. The EGR valve should close quickly. If the EGR valve closes slowly or not at all, replace the valve.

VACUUM AMPLIFIER DIAGNOSIS

On EGR systems with vacuum amplifiers, there are two types of amplifier design. Early-model units typically use a single connector, while late-model amplifiers have two connectors. To test the system:

1. With the engine at operating temperature, perform the throttle test. Slowly open and close the throttle while keeping the engine under 3000 rpm. If the EGR valve opens and closes, everything is okay. If the valve fails to open, proceed to the next step.
2. Pull the vacuum hose off the EGR valve and connect the hose to a vacuum gauge. Repeat the throttle test. If a strong vacuum signal (10 or more inches) is reaching the EGR valve, the amplifier is okay and the problem is a bad EGR valve diaphragm. If no vacuum or a weak vacuum is detected, move on to step 3.
3. If the system has a temperature vacuum switch or solenoid between the amplifier and EGR valve, pull the hose off the TVS that leads to the amplifier and connect your vacuum gauge. Repeat the throttle test. If vacuum is reaching the TVS, the amplifier is okay and the problem is in the switch. If still no vacuum or only a weak signal is detected, go to step 4.
4. Reconnect the TVS hose. Pull the vacuum hose from the carburetor venturi signal port. Temporarily plug the port and connect a hand vacuum pump to the signal hose. Start the engine and send a 1- to 3-inch vacuum signal through the line to the amplifier using your hand pump. If the engine starts to idle rough, indicating that the EGR valve is opening, the venturi vacuum port may be clogged or obstructed inside the carb. A thorough cleaning will be necessary to clear the path. If nothing happens and the vacuum supply line to the amplifier is okay, the amplifier is defective and should be replaced.

It should be noted that on some cars, the vacuum amplifier is calibrated to generate a small vacuum (2 inches or so) at all times, even when there is no carburetor venturi vacuum signal. This is to maintain vacuum in the system for quicker response. This small amount of vacuum is not enough to cause the EGR valve to open (most require at least 10 inches of vacuum), so do not think that the amplifier is defective. If you discover that full manifold vacuum is reaching the EGR valve at idle, the amplifier is leaking vacuum internally and should be replaced.

The thing to remember when troubleshooting EGR systems is that the basic principle of operation is the same regardless of how many components and gizmos are incorporated into the system. There should be no EGR at idle or when the engine is cold. And there should always be EGR in a warm engine under part-throttle operation. If the EGR valve is not working, start at the valve and trace backward to see why vacuum is not getting through. The same applies to those situations where vacuum is reaching the valve when it is not supposed to. The most likely cause here is a temperature vacuum switch or solenoid that is open all the time, or a defective vacuum amplifier.

One other thing that should be checked when inspecting any EGR system is to make sure that all the vacuum plumbing is correctly routed. With all the vacuum hoses under the hood, it is easy to get things mixed up. If the hoses or ports are not labeled, you should refer to a good shop manual for vacuum routing diagrams.

9

How Carburetion Affects
Emissions

Nothing, except perhaps ignition misfiring, affects the composition of the gases that an engine pumps into its exhaust manifold more than the carburetor. Anything from a simple maladjustment of the mixture screw or a lagging choke to a more complex problem like a leaky power valve can send HC and CO emissions so far out of line that the car becomes a gross polluter.

The automakers recognized this very early in their fight for clean air, so they focused a great deal of their attention on the carburetor. Some of the modifications they made were simple and easy to live with, while others caused drivability problems—hesitation, stalling, rough idle, and so on—but all were aimed at providing the engine with a blend of air and fuel that would be burned as completely as possible inside the cylinders. This meant that the mixture had to be leaner in all modes of operation than what was fed to power plants of the pre-emission-control-era.

Before we go on, it would be a good idea to define some basic principles and terms that are related to the science of carburetion. To begin with, an engine always burns gasoline at a ratio of air to fuel of 14.7:1 by weight regardless of the ratio that is drawn into the cylinders. In other words, the chemical reaction itself, the rapid oxidation of gasoline, the fire in the cylinders, takes 14.7 pounds of air to 1 pound of fuel. Note that we said by weight, not volume. Since air is a whole lot lighter than gasoline, it takes a huge volume of it to oxidize the fuel—9000 gallons to 1 gallon of gas.

Engineers have a very arcane name for this sacred ratio. They call it *stoichiometric*. As we said earlier, this is the ideal, theoretical ratio, but anything from about 8:1 to 18:1 will run an engine. With a rich mixture (a larger amount of fuel to the same amount of air), there is more gasoline than necessary, so some of it will not find any oxygen to combine with and will just be pumped raw out of the exhaust pipe. With a lean blend, on the other hand, there is an overabundance of air, so all the fuel is consumed. That is, as long as it is not so lean that it misfires.

To put it another way, the reaction always takes place at the stoichiometric ratio no matter whether the carburetor or fuel injection system is supplying 8:1 or 18:1. It should be obvious that in terms of emissions control and fuel efficiency, the leaner you can get the mixture and still have it fire dependably, the better. That is what Chrysler's Lean Burn, Honda's CVCC, and Ford's now-deceased PROCO program are all about—getting that thin blend to burn.

BOILING

Now for some physics. Maybe you have never thought about it, but it is still true that only fuel vapor will burn, so gasoline has to change its state from a liquid to a gas or the engine will not run. To do this, it must absorb enough heat to boil.

You may very well be wondering how it can possibly boil when the engine is cold, and that is a good question. The answer is simple: Just as water boils at less than 212 °F at high altitudes because of the lower atmospheric pressure on the liquid, so the vacuum in a carburetor's venturi and intake manifold helps to boil gasoline. In other words, the boiling point is reduced enough to vaporize the atomized droplets that enter the intake stream from the carburetor even if it is way below zero outside. Until the engine warms up, however, only a small part of the gasoline that is inducted into the intake manifold will turn to vapor, so the choke is needed to provide a terrifically rich mixture, and the fuel that remains in a liquid state is just wasted. That is why it is so easy for an engine to fail a government emissions certification test during the very first part of the cycle.

BOOSTING

Any time that air is forced through a tube, a pressure drop occurs, and this phenomenon is what gets gasoline to move into a carburetor's throat. The speed of the passing column of air determines the strength of the vacuum, so, especially at low rpm, it is necessary to give it a boost. That is what the venturi is for. The principle is that whenever a restriction is placed in a tube, air rushing past it must speed up, and this acceleration causes an extra pressure drop that helps atmospheric pressure in the bowl force enough fuel out of the carburetor to result in a burnable blend. Also, this additional vacuum makes the gasoline more enthusiastic about vaporizing.

Then there is atomization. You would be unhappy if you pushed the button of a household aerosol can and the contents came out in a solid stream. Well, an engine does not like it much either when it gets fuel in a hard-to-digest, unbroken form, so one of the carburetor's functions is to help smash gasoline into tiny droplets.

These expose a lot more area to the air than a stream would, which aids in vaporization. And a mist or vapor has less inertia than big chunks of liquid, so it negotiates the curves in the intake manifold more easily.

PUTTING THEORY INTO PRACTICE

All carburetors depend basically on the theory described above, and if you are familiar with it, no specimen you are apt to encounter will be able to totally mystify you—provided that you also understand the systems that put it into practice.

RESERVOIR

A carburetor needs a reservoir of constant depth to supply its circuits evenly, and that, of course, is the bowl. Most of the time, the engine cannot possibly use all the gasoline the fuel pump can supply, so the float and needle and seat setup stops the flow (Fig. 9-1). It may not seem possible that a tiny metal pontoon or a little piece of plastic foam could have enough buoyancy to shut down the 5 or 7 pounds per square inch (psi) the pump puts out, but leverage gives it plenty of strength to do the job.

So it is not the fuel pump that forces gas into the intake stream. Atmospheric pressure does it by bearing down on the surface of the liquid in the bowl. Naturally, there has to be a vent to let the weight of all those miles of air in. In the old days, this was right on top of the bowl, but modern carburetors have it in the form of a tube that sticks up out of the throat and takes advantage of the ram effect of the moving air column to push on the gasoline.

Now we come to a point that is in some dispute among experts: Does a dirty air filter element richen the mixture and reduce fuel mileage? The conventional view is that it does. But there are some dissenting voices based on the fact that since the

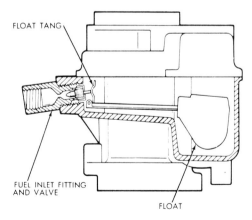

Figure 9-1 The float and needle and seat valve system keeps fuel in the bowl at the proper level. Leverage gives the float enough force to overcome fuel pump pressure. (Chrysler)

main bowl vent is inside the air cleaner, it is subject to the same pressure drop due to a clogged filter to which the rest of the intake stream is subject. That means that the force of the atmosphere on the fuel in the bowl is reduced, so less gasoline is pushed through the jets. We have heard of dynamometer tests where the blend could not be made richer even by taping up almost all of the air filter element. Maximum available power was certainly down, but not fuel mileage. According to the engineers who conducted these experiments, the clogged-filter-equals-rich-running idea is a leftover from the days of external bowl vents.

Our opinion? We will be politic and say that it is academic—in the real world, a dirty filter should simply be replaced, eliminating the whole question.

At idle, the velocity of the air entering the carburetor and hence the vacuum generated is too low to get fuel moving through the cruising system, so it has to come in somewhere else. That is the function of the idle circuit (Fig. 9-2), which is made up of a port below the throttle plate where there is lots of vacuum, a passage from the bowl with little air bleed holes in it to aid in atomization, and an adjustment screw.

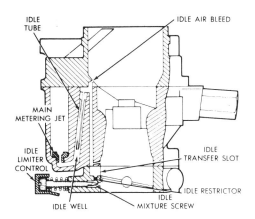

Figure 9-2   The idle circuit is controlled by the mixture screw. The transfer or off-idle slot or ports admit extra fuel as the throttle opens and expose them to vacuum. (Chrysler)

Other, similar outlets are used to provide a smooth transition from idle to moderate rpm, and these are called *transfer* or *off-idle ports*. They are positioned higher up in the throat than the idle ports and are progressively uncovered and exposed to vacuum as the throttle plate opens. In most specimens, they get their supply of fuel from the same tube as the idle port, but they are not affected by the mixture screw.

The cruising or main metering system is the next to come into action (Fig. 9-3). Its nozzle projects into the part of the venturi where there is the highest vacuum, and the maximum amount of gas it can spray is controlled by the diameter of the main jet through which it is supplied. This is the most efficient circuit in a carburetor, usually calibrated to provide the ideal 14.7:1 fuel/air ratio. It starts working before the last off idle port is uncovered, then takes over completely.

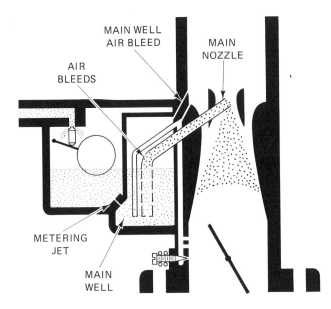

Figure 9-3 The cruising circuit gets its supply of fuel from the main well. The amount is controlled by the size of the hole in the metering jet. (Chevrolet)

PEDAL TO THE METAL

During hard acceleration, when the throttle is open wide (Fig. 9–4), far too much air enters the carburetor for the cruising system to handle. Without help, it would allow the mix to lean out, and acceleration and top speed would be way below what the engine is actually capable of producing.

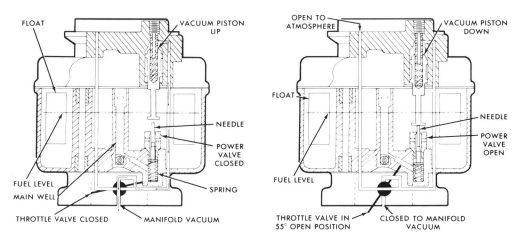

Figure 9-4 When the throttle approaches its wide open position, vacuum drops. This puts the power enrichment system into action. High vacuum holds the piston up, but the low vacuum of heavy acceleration lets it fall, opening the power valve and admitting extra fuel. (Chrysler)

Enter the power circuit (Fig. 9–5), which opens up and allows extra gasoline to pass through the carburetor. It can be either a separate fuel valve, or a metering or step-up rod that normally blocks some of the main jet's flow, but is pulled up out of the way mechanically or by vacuum when the engine needs more fuel. The vacuum-operated type commonly opens at 5 to 7 in. Hg.

You could drive a car that was supplied by just the circuits described above, but it would not be much fun. If you stepped on the gas just a little too rapidly, a blast of wind would rush in and put out the fire before enough fuel could get moving to make a flammable mixture.

So carburetors have an accelerator pump to compensate for gasoline's inertia (Fig. 9–6). It squirts an extra shot of fuel into the intake stream whenever the throt-

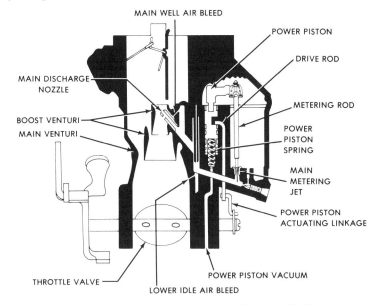

Figure 9-5   This type of power system raises a metering rod under low vacuum, increasing the main metering jet's flow. (Buick)

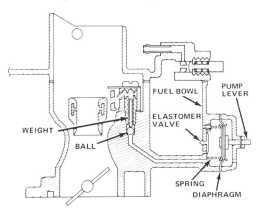

Figure 9-6   The accelerator pump squirts extra gasoline into the carburetor's throat whenever the throttle is opened. (American Motors Corp.)

tle is opened and, since it works mechanically, does it before the engine gets a chance to choke on those big chunks of plain air. It is a simple piston pump with one-way inlet and outlet valves and an air bleed, weight, or spring arrangement that keeps vacuum from drawing fuel through the nozzle while idling or cruising.

SUPER RICH

When the engine and the weather are both cold, gasoline is very reluctant to vaporize, so an extremely rich mixture is needed to get that blaze going in the cylinders. And that is where the choke comes in (Fig. 9-7). It closes off the mouth of the carburetor so that fuel is pumped through every possible orifice as the starter is cranked.

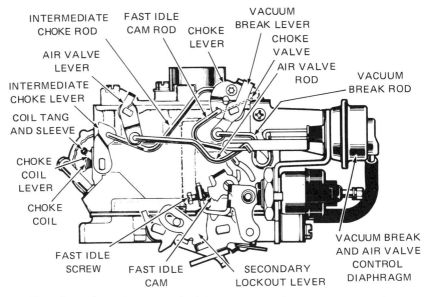

Figure 9-7 The modern automatic choke system comprises many parts to get it to perform efficiently. (American Motors Corp.)

Once the power plant fires up, vacuum pulls on a diaphragm or piston (known as the vacuum break or choke pull-off), which opens the choke a crack. Then a thermostatic coil, which is heated by exhaust or electricity (Figs. 9-8 and 9-9), continues to move the plate until it no longer impedes airflow.

Finally, there is the hot idle compensator that is used on many carburetors to make up for the difference in density between warm and cool air (Fig. 9-10). On a hot summer day, the air that gets past the throttle plate at idle is so thin that the mixture becomes excessively rich. The compensator is simply an extra air passage that opens up to bring the blend back into line. It is controlled by a tiny bimetal valve.

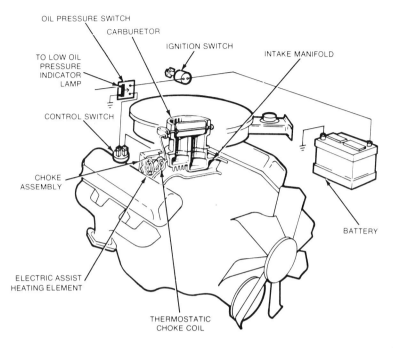

OIL PRESSURE SWITCH

CARBURETOR

IGNITION SWITCH

INTAKE MANIFOLD

TO LOW OIL PRESSURE INDICATOR LAMP

CONTROL SWITCH

CHOKE ASSEMBLY

BATTERY

ELECTRIC ASSIST HEATING ELEMENT

THERMOSTATIC CHOKE COIL

Figure 9-8 Many choke designs today incorporate electric heating to expand the thermostatic coil faster and so open the choke more rapidly. (Chrysler)

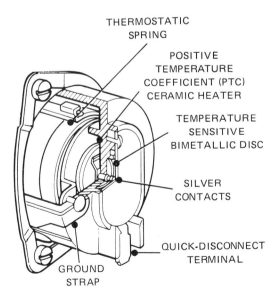

THERMOSTATIC SPRING

POSITIVE TEMPERATURE COEFFICIENT (PTC) CERAMIC HEATER

TEMPERATURE SENSITIVE BIMETALLIC DISC

SILVER CONTACTS

QUICK-DISCONNECT TERMINAL

GROUND STRAP

Figure 9-9 This is a cutaway of an electric assist choke coil. (Ford)

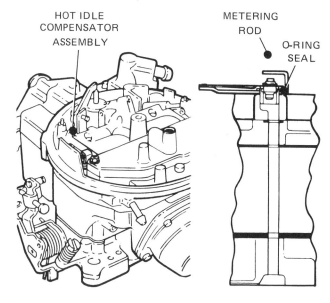

Figure 9-10 The hot idle compensator opens to admit extra air which keeps the mixture from getting too rich under hot conditions when the air is thin. (Ford)

SUPERCARB

Then there are the new systems that control the air/fuel ratio much more closely than any ordinary carb has ever been able to do, but without resorting to fuel injection. All four of our domestic automakers now have electronic feedback or closed-loop carburetion setups (Fig. 9–11).

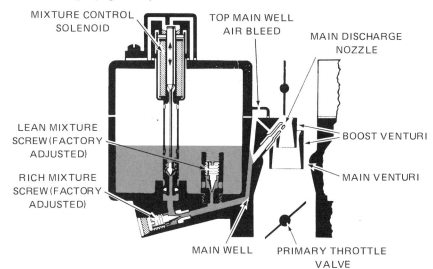

Figure 9-11 This feedback/closed-loop carburetor uses a mixture control solenoid to keep the air/fuel ratio as nearly perfect as possible. The solenoid is operated by an electronic control unit. (American Motors Corp.)

TRIPLE CAT

The basic reason for the introduction of this concept is the three-way or reduction catalyst, the kind that contains platinum and rhodium so that it burns up No_x along with HC and CO. This extra reaction is very fragile. It takes place only if the amount of oxygen in the exhaust stream is exactly right, the result of a perfect 14.6:1 air/fuel ratio, something the old-fashioned carburetor just cannot guarantee.

GM's Computer Controlled Catalytic Converter (C-4 for short, called C-3 in later models for "Computer Command Control") system first appeared on 1978 2.5-liter Pontiac Sunbirds built for California. It uses a special carburetor, an Electronic Control Module (ECM), and oxygen, temperature, vacuum, throttle position, and rpm sensors to adjust the air/fuel ratio automatically.

The carburetor (E2SE, E2ME, or E4ME) has a solenoid inside that pushes down a metering rod when it is energized by the ECM, and this reduces the flow of fuel to the main metering and idle circuits. Also, when the rod is down, it opens an auxiliary air bleed to the idle system, further leaning the mixture.

The solenoid cycles 10 times per second and the amount of "on" time relative to "off" time is what determines the richness of the blend.

MAGICAL MYSTERY SENSOR

The most important input the ECM gets comes from the zirconia oxygen sensor that is screwed into the exhaust manifold so that the hot gases flow over its business end. This device generates voltage that varies with the amount of 0_2 in the exhaust. As the oxygen content falls (indicating a rich mixture), the sensor sends the ECM more volts. The computer converts this rise into increased carburetor solenoid "on" time, which leans the mixture down.

To maintain good idle and drivability, the ECM takes signals from the other sensors into account also and modifies its commands to the carburetor accordingly.

FORD FEEDBACK

The 1980 Ford System has numerous variations. First, there is the one for California 2.3-liter fours that uses a Holley/Weber 6500 staged two-barrel (a modified 5200). Next is the 7200 VV (Variable Venturi) Motorcraft carb Air Bleed system used in 5.8-liter power plants made for California. Finally, there is the 7200 VV back-suction type that is used on 5.8-liter engines in 49 states, and 4.2- and 5.0-liter California V8's.

Of course, all Fords that are equipped with feedback carburetion have a microprocessor for brain power, and it needs the intelligence collected by several spies under the hood to make decisions about air/fuel ratio strategy. These include sensors for rpm, vacuum, temperature, and throttle position, and an exhaust gas oxygen (EGO) sensor basically similar to the GM unit.

BACK SUCTION

All these carbs use different methods of regulating the fuel that passes through them so that the catalyst can do its job efficiently. We will describe the back-suction type.

The carburetor—essentially a 2700 VV—has an actuator stepper motor (Fig. 9-12) mounted on its right side that moves a vacuum metering rod according to commands from the EEC (Electronic Engine Control) III computer. The position of the metering rod in its orifice regulates the amount of vacuum that is applied to the fuel bowl. The larger the opening, the greater the vacuum and the leaner the mix because it is harder for the gasoline to leave the bowl.

The motor has 100 steps within its 0.4-inch range and four separate coil windings. The microprocessor energizes the windings sequentially to put the metering rod where it wants it.

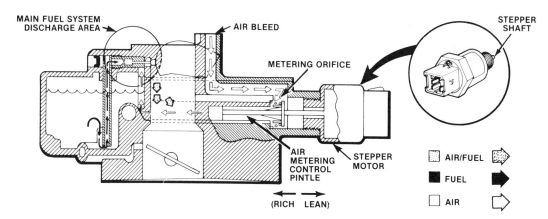

Figure 9-12 Instead of controlling the mixture by opening and closing a jet, this feedback carburetor uses a stepper motor that varies the amount of atmospheric pressure that bears on the fuel in the bowl. (Ford)

SELF-CRITICISM

You might think that a mechanic would need an advanced electronics engineering degree to be able to troubleshoot feedback carburetion. And you would be right except that the manufacturers foresaw the potential service problems they were unleashing on an unwary industry, so they built some amazing self-diagnostic abilities into their systems. Because there are so many complicated variations, you will need the specific service information for the car at hand in order to take advantage of the on-board troubleshooting programs.

NOT MUCH TO DO

Besides setting the choke and fast and curb idle speeds, the only adjustment that is normally available on a carburetor is that of idle mixture. Since the early 1970s, idle mixture screws have had limiter caps installed over them at the factory. These limit the amount the screws can be turned, thereby keeping the blend within acceptable parameters. These are frequently removed during service, usually to make the setting richer, but a reasonably smooth idle should be obtainable within their limits.

On later models, especially those with feedback/closed-loop systems, the mixture screws are much more tamper-proof. For example, they may be concealed under a plug that must be drilled out. Unless you have the proper equipment and service information, it is folly to attempt to make idle mixture adjustments in ordinary service.

Setting the idle mixture today requires either an infrared exhaust gas analyzer (an expensive piece of equipment that tells you the amount of HC and CO that is in the exhaust), or the use of a propane enrichment procedure.

EXHAUST ANALYSIS

The infrared analyzer is useful not only in adjusting the idle mixture, but also in troubleshooting carburetion, ignition, and other problems. The meters on an exhaust analyzer's face are labeled "Hydrocarbons (HC)" and "Carbon Monoxide (CO)," but a brief explanation of what these gases are is certainly appropriate here. (See Chapter 12 for more information on emissions testing.)

HC is simply unburned gasoline and oil that is present in an engine's exhaust because of incomplete combustion. Even the best engine in the world will pump some HC because the flame front is quenched when it hits the relatively cool cylinder walls, leaving a small amount of fuel unburned. The analyzer reads this trace as n-hexane in parts per million (ppm).

Abnormally high levels of HC are usually caused by an idle mixture that is too rich or one that is so lean that it causes misfiring, a leaky needle and seat or a float problem, trouble in the power or main metering circuit, plug misfiring due to fouling or a bad secondary ignition component, a malfunctioning air injection pump or other emission control device, or anything that lets engine oil into the fire; namely, worn-out rings, valve guides, or stem seals. (See Chapter 13 for more information on emissions troubleshooting).

CO-CONSPIRATOR

HC's deadly partner, CO, is a by-product of combustion in the presence of too little oxygen—if there were sufficient O_2, harmless CO_2 would be generated—so any condition that restricts air intake or causes extra fuel to be drawn into the engine can produce excessive carbon monoxide. These include a rich idle mixture or choke setting, too much timing advance, or low idle speed (both of which keep the throttle plates closed more than they should be hence cutting down on air ingestion), a leaking power valve or accelerator pump, a clogged PCV system, an inoperative air injection pump or other smog control contraption, and a clogged air filter. CO, by the way, is measured in percentage by volume.

PROPANE CONNECTION

Propane enrichment idle mixture setting procedures were adopted to meet emissions control laws. All that is needed is a propane tank, valve, and hose setup and the proper information for the specific car. Basically, the adjustment is made as follows:

1. Warm up the engine and disconnect the PCV hose and the charcoal canister purge hose from the air cleaner, and plug the air cleaner nipples; or remove the air cleaner.
2. With the engine idling, hook up a tachometer and insert the propane hose into the air cleaner snorkel or attach it to one of the air cleaner nipples. If the air cleaner is removed, connect the propane hose to the choke pull-off vacuum nipple.
3. Open the propane valve slowly until maximum rpm is reached (keep the propane bottle vertical).
4. Turn the carb idle speed screw or solenoid plunger until you have achieved the specified propane-enriched rpm.
5. Continue to open the propane valve. If the speed rises, let it do so until it levels off. Then reset the curb idle speed screw to the enriched rpm.
6. Shut off the propane valve and turn the idle mixture screw until the smoothest idle at the specified hot idle speed is attained.

After reading this chapter, it should be obvious to you that carburetion has a profound effect on emissions. If the engine is fed a mixture that is too rich, or so lean that it causes misfiring, an unacceptable level of pollutants will escape from the tailpipe.

Although this chapter was about carburetion, we should mention that there is a discussion of fuel injection in Chapter 15.

10

How Ignition Affects
Emissions

In Chapter 9 we discussed how carburetion and the air/fuel mixture affect emissions. In this chapter we examine how burning the air/fuel mixture inside the engine as determined by ignition timing affects emissions.

FOUR-STROKE REVIEW

To understand the how's and why's of ignition timing, we need to review briefly how an internal combustion engine works (see Fig. 10–1). There are four basic strokes that an engine goes through to produce power: intake, compression, power, and exhaust. On the intake stroke, the intake valve opens as the piston begins to travel downward in the cylinder. The piston creates a vacuum and pulls a fresh charge of air and fuel into the cylinder. When the piston reaches the bottom of its travel, it reverses direction and starts back up. This is the beginning of the compression stroke. The intake valve closes so that the air and fuel will not be pushed back out of the cylinder. As the piston continues its upward travel, it compresses the air and fuel. This helps to mix the air and fuel more thoroughly and raises the temperature of the mixture so that it will burn more readily. As the piston reaches the top of its travel, the fuel mixture is ignited by a spark. This starts the mixture burning, which produces heat and pressure. The hot expanding gases push the piston down while they continue to burn. This is the power stroke. As the piston reaches the bottom of its travel and starts back up again, the exhaust valve opens so that the piston can shove the exhaust gases out of the cylinder. This is the exhaust stroke. The exhaust valve closes as the piston reaches the top of the cylinder (called top dead center, or

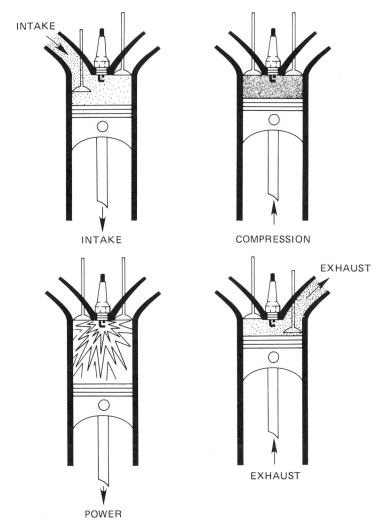

Figure 10-1   Basic 4-stroke cycle of the internal combustion engine.

TDC). The intake valve then opens as the piston starts back down again to begin the process all over.

If all this has you confused, think of it this way: the crankshaft must make two full revolutions for the piston to complete the four-stroke cycle. On the first revolution, the piston does the intake and compression strokes. On the second revolution, it completes the power and exhaust strokes. This means that ignition occurs every other time the piston passes the top-dead-center position.

When an engine is idling at 600 rpm, the crankshaft is spinning around 600 times a minute or 10 times a second. Each piston is also reciprocating up and down at the same 10 times a second rate. Since combustion occurs only on the power stroke every other revolution, ignition takes place five times a second at 600 rpm.

The air/fuel mixture has plenty of time to burn at idle because the duration of the power stroke at 600 rpm is on the order of 1/20 (0.05) second. The actual combustion of the air/fuel mixture takes place in about 1/200 (0.005) second, so you can see that there is plenty of time for the fuel to burn and expand before it is pushed out of the cylinder on the exhaust stroke.

At highway speeds, a typical engine might be turning 3000 rpm. At this speed, the pistons will be traveling up and down at the rate of 50 times a second. With ignition taking place every other revolution, combustion occurs 25 times a second. The time available during the power stroke at 3000 rpm is now on the order of 1/100 (0.01) second—still enough time for the fuel to complete burning on the power stroke but not enough time for cylinder pressures to reach a maximum at the right point in the power stroke.

For an engine to deliver maximum fuel economy and power, cylinder pressure should reach a maximum early in the power stroke while the piston is accelerating downward. If the instant of maximum pressure occurs too late in the power stroke, it will not produce as much horsepower. And if maximum pressures are not achieved until very late in the power stroke, some of the oomph will be lost out the exhaust valve when the exhaust stroke begins. Timing, therefore, plays a very important role in both power output and economy.

To give the air/fuel mixture sufficient time to burn so that maximum cylinder pressures can be achieved at the best point in the power stroke, the ignition timing is advanced before top dead center. Instead of happening exactly at TDC as one might assume, ignition takes place farther and farther before TDC as engine speed increases. In other words, ignition takes place toward the end of the compression stroke.

For example, a typical engine at idle may be timed at anywhere from 0°TDC to 10°BTDC (Fig. 10–2). Some may even be timed several degrees after top dead center. Remember that at idle there is plenty of time for combustion, so little or no initial advance is needed. At 3000 rpm, however, considerable advance is needed to achieve maximum cylinder pressure early in the power stroke. Such an engine might have 28 to 34° ignition advance, the amount of advance increasing in proportion to engine speed.

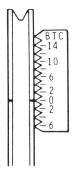

Figure 10–2 Ignition timing marks can be found on the crankshaft pulley, harmonic balancer or flywheel. (Chevrolet)

TIMING AND DETONATION

The higher the rpm, the more timing advance an engine needs to maximize power output and fuel economy. The timing must advance in proportion to the rpm rate. If the fuel is ignited too far in advance, however, the pressure of the expanding gases rises too quickly and peaks before the piston can respond. This causes detonation and increases HC emissions.

Think of combustion as an expanding balloon (Fig. 10–3). Under normal circumstances, the flame front expands and fills the combustion chamber outward from the spark plug. When there is too much advance, though, the rapidly increasing pressure inside the combustion chamber causes fuel to ignite spontaneously in other areas of the chamber. This is akin to several balloons expanding all at once. And when the flame fronts collide, they do so with great force. This produces the sharp knock or ping noise that is characteristic of detonation. Under severe circumstances, detonation can crack or punch holes in pistons, crack heads and piston rings, flatten connecting rod bearings, and eat away at head gaskets. Hydrocarbon emissions are also increased because the erratic combustion can leave little pockets of unburned fuel.

For optimum performance and power, an engine should be timed so that it is just on the verge of detonation. The only drawback to this approach is that an engine's detonation resistance changes. It can vary with the quality of gasoline being used or with the weather. Brand X regular-grade gasoline may have a lower octane rating than brand Y, and thus a greater tendency to detonate. High relative humidity increases the effective octane rating of gasoline, whereas dry weather does just the opposite. What this means to real-life driving conditions is that an engine that is tuned for optimum performance on a rainy day may tend to knock and ping during dry weather. To play it safe, therefore, the manufacturer's recommended timing settings have a built-in safety margin to account for fuel and weather differences. Late-model cars with computerized engine controls can push the limits somewhat if they are equipped with a knock sensor. This device picks up pressure pulses in the intake manifold that are produced when detonation occurs. The sensor tells the computer

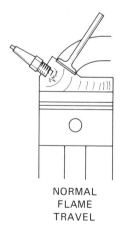

NORMAL
FLAME
TRAVEL

DETONATION

Figure 10–3 During normal combustion, the flame front travels outward from the spark plug like an expanding balloon. But when cylinder pressures exceed the octane rating of the fuel, spontaneous combustion can create multiple flame fronts. When the flame fronts collide, they produce a sharp knock or rapping noise.

the timing is overadvanced for conditions, so the computer backs off the timing a couple of degrees.

IGNITION BASICS

Now that we have covered the basic operating principles of a four-stroke internal combustion engine, let's define some of the terminology that is used to describe ignition timing:

1. If ignition occurs exactly at top dead center, there is 0° of timing advance.
2. If ignition occurs *after* TDC, the timing is said to be *retarded*.
3. If ignition occurs *before* TDC, the timing is said to be *advanced*.
4. *Initial timing* is the amount of advance or retard the ignition has at idle as set in accordance with the manufacturer's instructions. For most vehicles, this means disconnecting the vacuum advance hose from the distributor or computer. Initial timing can also be called *basic timing*.
5. Increasing the amount of ignition advance is called *advancing* the timing. Decreasing the amount of advance is called *retarding* the timing.

IGNITION CURVES

As described earlier, the amount of timing advance must increase in proportion to engine speed. This is accomplished in two ways: mechanically or electronically.

In distributors with centrifugal advance mechanisms, two small flyweights control the rate of advance (Figs. 10–4 and 10–5). The weights are located in the distributor

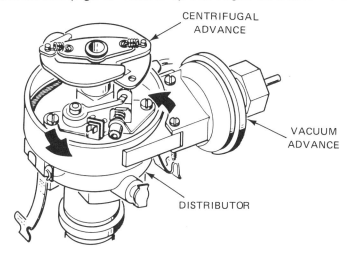

Figure 10-4   The centrifugal and vacuum advance mechanisms on the distributor. Many engines with computerized emissions control systems do not have these devices. (Chevrolet)

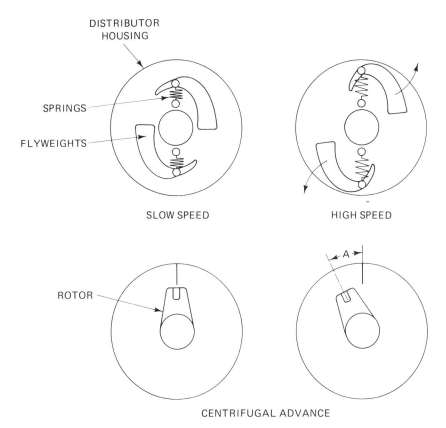

Figure 10-5 Centrifugal advance is controlled by two flyweights and springs. As engine rpm's increase, the weights are thrown outward against spring tension. This moves the rotor and trigger wheel (electronic ignitions) or breaker cam (point ignitions) into an advanced position.

under the rotor or under the breaker point or pickup plate. As the distributor spins faster and faster with increasing rpm, the flyweights are thrown outward by centrifugal force. The size of the weights and the strength of the springs that resist the weights determine the rate of advance. Changing the weights and/or springs will change the advance curve (Fig. 10–6). The mechanism is connected to the distributor rotor shaft so that the outward movement of the weights causes the shaft to rotate slightly in the direction opposite of normal rotation. This advances the position of the cam plate or pickup trigger wheel, which in turn signals the coil to fire the plugs sooner.

On cars equipped with computerized engine controls, the computer calculates the amount of ''centrifugal'' advance the engine needs based on rpm. The advance curve is programmed into the computer, and may be varied by other sensor input. There would be no flyweights or advance mechanism in the distributor on such engines.

To improve fuel economy, most engines also employ vacuum advance (Figs. 10–7 and 10–8). Vacuum advance differs from centrifugal advance in that it is engine-

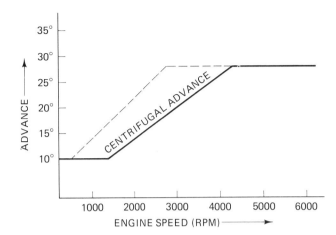

Figure 10-6 The rate at which centrifugal advance occurs determines the "advance curve" as shown by the solid line. The amount of advance increases as rpm's increase. Installing heavier weights or weaker springs will change the curve to the dotted line. Now centrifugal advance will occur much more rapidly and at a lower rpm.

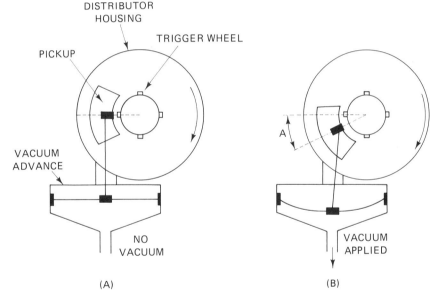

Figure 10-7 Vacuum advance only works when vacuum is applied to the diaphragm. In (a), there is no vacuum advance. In (b), vacuum has pulled the diaphragm outward. This causes the pickup or points to slide into an advanced position with respect to the trigger wheel or breaker cam inside the distributor.

load sensitive. It increases when there is a light load on the engine and decreases when the engine is run at wide-open-throttle positions. Centrifugal advance, on the other hand, is speed sensitive (Fig. 10-9). It increases with rpm.

The vacuum advance mechanism itself is fairly simple. A vacuum hose from the carburetor or intake manifold is connected to a vacuum diaphragm on the distributor. The diaphragm moves the breaker point or pickup plate to advance the

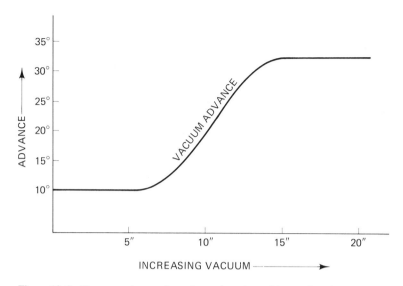

Figure 10-8 Vacuum advance depends on throttle position and engine load, not rpm's.

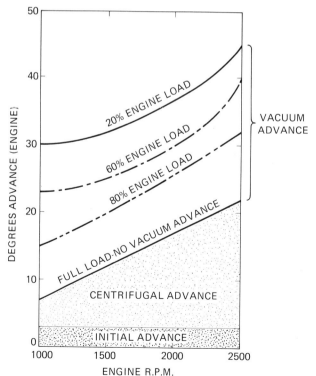

Figure 10-9 Ignition advance under various operating conditions. The total amount of advance is initial plus centrifugal plus vacuum.

timing. Many Fords have a dual vacuum diaphragm that both advances and retards timing. On late model Chryslers with Lean Burn, a vacuum diaphragm on the computer reads engine vacuum so that the computer can calculate timing advance as needed. The advance curve is programmed into the computer and can be modified by other sensor input.

Under light load, an engine can handle more advance without knocking. The benefit of adding additional timing advance under such circumstances is better mileage and more pep. Under heavy load, however, cylinder pressures increase rapidly and the tendency to detonate rises. To compensate for the load, the additional timing advance must be temporarily reduced. Since engine vacuum drops in proportion to the load applied, using vacuum to control timing advance allows the timing to respond to the conditions.

At idle (Fig. 10-10), light load, and during deceleration, intake manifold vacuum is very high. Therefore, a simple vacuum advance mechanism would provide additional timing advance under these circumstances. As the throttle is opened wider (Fig. 10-11), intake manifold vacuum drops. At full throttle (Fig. 10-12), there is very little manifold vacuum. Under such conditions, the vacuum advance mechanism would not supply any additional timing advance.

The vacuum signal for the vacuum advance comes from one of three sources: intake manifold, ported vacuum, or venturi vacuum. With intake manifold vacuum, the vacuum hose is simply connected to a fitting on the manifold or the base of

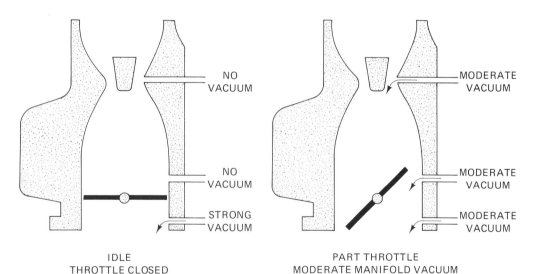

IDLE
THROTTLE CLOSED
HIGH MANIFOLD VACUUM

PART THROTTLE
MODERATE MANIFOLD VACUUM

Figure 10-10 At idle, the throttle plate is closed. There is a strong vacuum in the intake manifold, but no vacuum at the ported or venturi connections.

Figure 10-11 When the throttle is opened, air begins to flow down through the venturi. This produces a venturi vacuum signal. The ported vacuum opening is also exposed so a vacuum signal can now be read there as well.

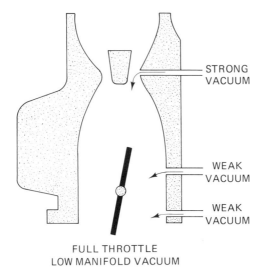

STRONG VACUUM

WEAK VACUUM

WEAK VACUUM

FULL THROTTLE
LOW MANIFOLD VACUUM

Figure 10-12 At wide open throttle, overall vacuum in the engine is low so both intake and ported vacuum signals will be weak. However, air flow through the venturi is at a peak, so venturi vacuum will be strong.

the carburetor. With ported vacuum, the hose is connected to a fitting on the carb which vents above the throttle plates. At idle, there is no vacuum signal because the port is above the throttle plates (which are closed). As the throttle is opened, the port is exposed to intake vacuum. The vacuum signal passes through the hose and timing is advanced. On some vehicles, the vacuum hose is connected to a fitting on the carb that vents into the venturi. Reading engine vacuum at this point produces a faster response, but the venturi vacuum signal is typically too weak to move the distributor diaphragm. The vacuum hose from the carb, therefore, is usually connected to a gadget called a *vacuum amplifier*. The amplifier reads the venturi vacuum signal and responds by allowing intake manifold to pass through a valve to the distributor advance diaphragm.

TIMING AND EMISSIONS

Engineers discovered that ignition timing has a significant impact on tailpipe emissions. Generally speaking, they found that retarding the ignition timing at idle and during deceleration reduced hydrocarbon emissions. When the timing is retarded, combustion occurs later in the power stroke. This increases exhaust gas temperatures and promotes more complete burning of hydrocarbons in the exhaust system. When the hot exhaust gases enter the exhaust manifold and meet fresh oxygen as supplied by the air injector reactor (AIR) system, the unburned HC continues to burn. Add a catalytic converter to accelerate the process, and the result is almost complete combustion of any hydrocarbons that were not already burned inside the engine. Retarded ignition timing also requires a slightly wider throttle opening. This is necessary to increase the flow of air and fuel so that the idle speed can be maintained. The wider opening and increased flow promote better mixing of air and fuel. This aids combustion and reduces HC emissions.

CONTROLLING EMISSIONS

The easiest way to retard ignition timing during idle and deceleration to reduce emissions is to use ported vacuum advance (Fig. 10–13). At idle there is no vacuum advance because the port is located above the throttle plates. During deceleration there is no vacuum advance because again the throttle is closed.

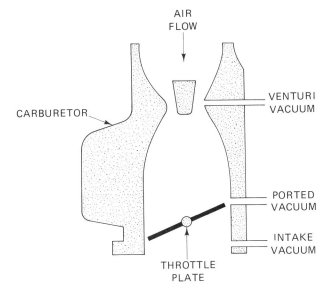

Figure 10-13 Three sources of possible vacuum signals for vacuum advance. Intake vacuum is taken below the throttle plate. Ported vacuum is just above the throttle, and only works when the throttle is opened. Venturi vacuum is taken at the narrowest part of the carburetor throat.

Another way to retard ignition timing during idle and closed throttle deceleration is to use a combination of ported and intake manifold vacuum. The two vacuum sources are balanced against one another through a spring-loaded valve. Manifold vacuum reaches the distributor only when ported vacuum is sufficient to open the valve. On Ford dual vacuum diaphragm distributors (Fig. 10–14), carburetor ported vacuum is connected to one side of the diaphragm and intake manifold vacuum to the other. At idle and during closed-throttle deceleration, ported vacuum is zero and intake manifold vacuum is high. The ported vacuum side of the diaphragm produces timing advance while the intake manifold vacuum side produces retard. Thus timing is retarded at idle and during deceleration when manifold vacuum is strongest, and advanced during other modes of operation when ported vacuum is strongest.

Many systems, including the Ford dual vacuum diaphragm unit just described, use ported vacuum switches (PVS) or temperature vacuum switches (TVS) to reduce emissions (Figs. 10–15 and 10–16). A PVS or TVS installed in the vacuum line between the distributor and vacuum source prevents vacuum from reaching the distributor during certain modes of operation. For example, tailpipe emissions are lower on some cars if vacuum advance is restricted to cruising speeds only. A PVS is used to prevent the vacuum signal from reaching the distributor until the transmission is shifted into high gear. This approach is known as *transmission-controlled spark* (Figs.

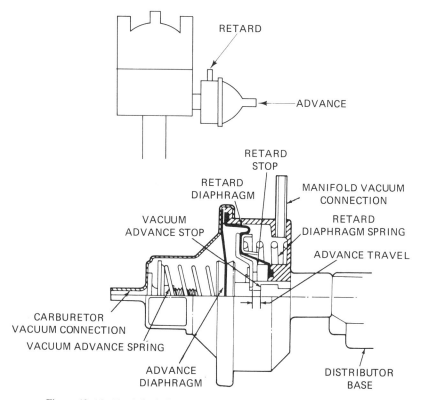

Figure 10-14 Ford dual diaphragm vacuum advance. (Ford Motor Co.)

10–17 and 10–18). A small electrical switch located on the shift linkage controls a ported vacuum solenoid. When the trans is put into high gear, the switch grounds and energizes the solenoid, allowing vacuum to pass to the distributor.

On many engines, emissions can be reduced if vacuum advance is blocked until the engine reaches operating temperature. This is because the relatively cold engine causes droplets of fuel to condense on the cylinder wall surfaces. This increases unburned hydrocarbon emissions so that higher exhaust temperatures are needed to burn the HC. A TVS is installed in the vacuum advance line to block vacuum until the engine reaches operating temperature. The TVS is screwed into the engine block or intake manifold so that the tip of the TVS is in contact with the engine's coolant. When the coolant reaches 180°F or so, the wax plug inside the TVS expands and opens the line between the vacuum source and the distributor.

Some TVSs work in just the opposite way. On some engines, vacuum advance is needed when the engine is cold to improve idle quality and performance. On such applications, the TVS is designed to allow full vacuum advance until the engine is warmed up. The TVS then closes off the intake manifold vacuum line and opens a ported vacuum line.

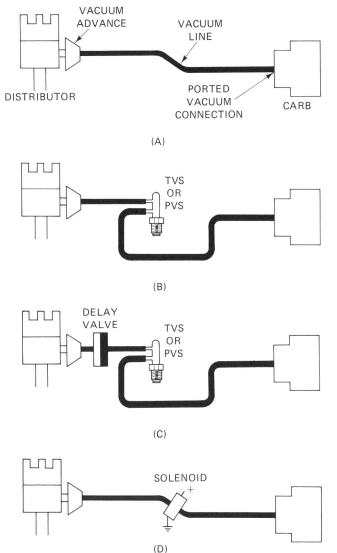

Figure 10-15 Methods of controlling vacuum advance. In (a), the vacuum advance is connected directly to the ported vacuum connection on the carburetor. When the throttle opens, the timing advances. In (b), a TVS or PVS is installed in the vacuum line to prevent advance until the engine is warmed up. In (c), a delay valve has been added to prevent the timing from advancing too quickly. In (d), an electrical solenoid is used to turn vacuum advance on and off. Such a system is used in transmission controlled spark systems. A solenoid can also be computer controlled so that vacuum advance only occurs under certain operating conditions.

On other engines, a TVS is used to perform yet another function. If the engine starts to overheat, the TVS opens up a line between intake manifold vacuum and the distributor temporarily to advance the timing. This increases idle speed, which in turn circulates coolant through the engine more quickly. As the engine cools back down to normal temperature, the TVS closes the intake manifold vacuum line and idle speed returns to normal. Such a switch is often called a *coolant temperature override* (CTO) switch.

Another device used to modify vacuum advance is the *spark delay valve* (Fig. 10-19). This device works like a restriction in the vacuum line. Depending on the

HOT-COOLANT OVERRIDE SWITCH

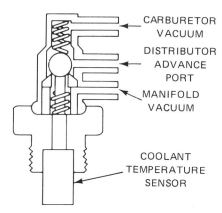

Figure 10-16 A ported vacuum switch. The switch blocks vacuum to the distributor from the carb until the engine is warm. If the engine overheats, manifold vacuum is fed to the distributor to advance timing. This causes the engine to run faster and cool down.

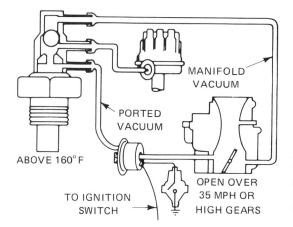

Figure 10-17 Schematic of how a typical transmission controlled spark system might be installed. The switch on the gear selector controls the solenoid in the vacuum line.

application, the valve can prevent full vacuum from reaching the distributor for a few seconds up to a half a minute or more. The delay valve is used to prevent sudden ignition advance that can cause combustion temperatures to soar, and thus increase NO_x formation. By delaying the advance for a short period of time and allowing it to build up gradually, peak combustion temperatures can be avoided and NO_x formation reduced.

Spark delay valves go by a variety of names but all do basically the same thing. Chrysler uses an orifice spark advance control (OSAC) valve (Fig. 10-20). Ported vacuum from the carburetor is routed to the OSAC valve, then to the distributor. When a vacuum signal reaches the OSAC valve, it takes about 20 seconds for it to pass through the valve and reach the distributor. General Motors uses a spark delay valve (SDV) or a vacuum delay valve (VDV) for the same purpose. The SDV or VDV is located in the vacuum line between the ported vacuum connection on the carburetor and a TVS. The valves have a 0.005-in. orifice restriction, which delays the vacuum signal from reaching the distributor by about 40 seconds.

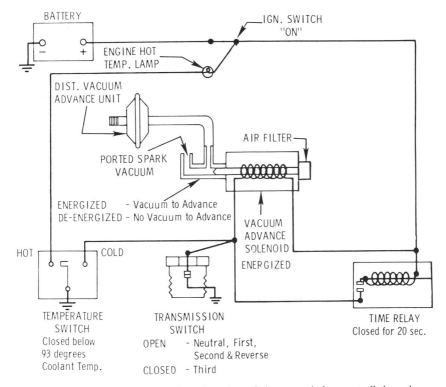

Figure 10–18 A more complicated version of the transmission controlled spark system. This one happens to be on an Oldsmobile. Like the system in Fig. 10–19, this one uses a temperature switch and a transmission switch. But it also has a time relay that provides 20 seconds of advance when the engine is first started to aid cold drivability. (Chevrolet)

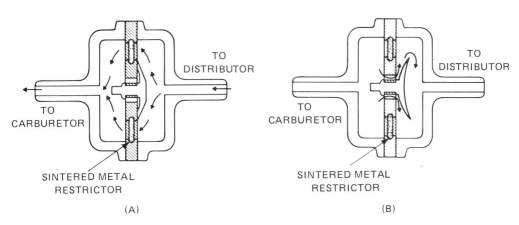

Figure 10–19 Spark delay valve. Depending on the amount of restriction, vacuum signals can be delayed from ½ second up to 60 seconds or more. In (a), the vacuum signal passes through the restrictor on its way to the distributor. If vacuum drops to zero (b), air can instantly rush back through a one-way valve. This allows the vacuum advance to drop instantly.

115

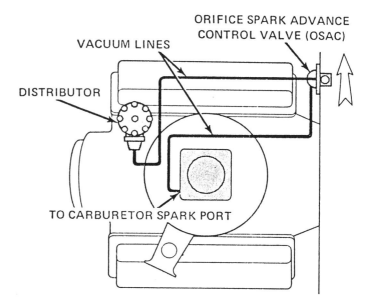

Figure 10-20   Chrysler OSAC type spark delay valve. (Chrysler)

Many spark delay valves use a porous metal filter to restrict the flow of vacuum. Such delay valves are also fitted with a one-way rubber valve. The valve allows vacuum to escape from the distributor side of the line when the ported vacuum signal drops to zero (as during deceleration or idle). In other words, the valves restrict vacuum one-way only. Therefore, it is very important that these valves be installed facing the right direction. Most are marked "DIST" on the distributor side or "CARB" on the carburetor side. They are also color coded according to the amount of delay they provide.

Spark delay valves can become clogged with dirt. If this happens, the delay period may be longer than usual, or there may be complete blockage of the vacuum signal. Delay valves can be tested with a hand-held vacuum pump. Apply vacuum to the carb side and see how long it takes for the reading to drop. Compare this to the manufacturer's specs to see whether or not the valve is doing the job it is supposed to do. When vacuum is applied to the distributor side, there should be no restriction and the reading should be zero.

ELECTRONIC SPARK CONTROL

The most sophisticated approach to controlling ignition timing for reduced emissions and optimum performance and economy is with a computer (Figs. 10–21 to 10–23). There are a variety of electronic spark controls, but all work on basically the same principle. Engine sensors tell the computer about rpm, coolant temperature, engine

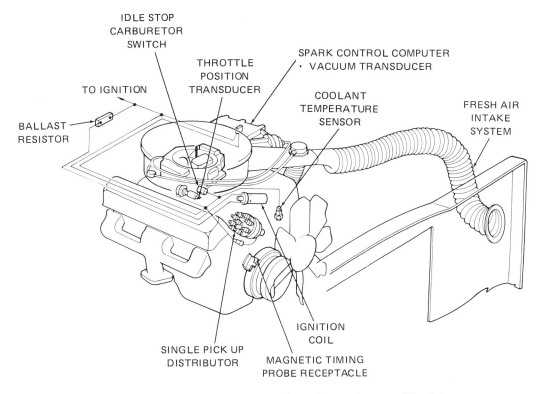

IDLE STOP
CARBURETOR
SWITCH

THROTTLE
POSITION
TRANSDUCER

SPARK CONTROL COMPUTER
· VACUUM TRANSDUCER

TO IGNITION

COOLANT
TEMPERATURE
SENSOR

FRESH AIR
INTAKE
SYSTEM

BALLAST
RESISTOR

IGNITION
COIL

SINGLE PICK UP
DISTRIBUTOR

MAGNETIC TIMING
PROBE RECEPTACLE

Figure 10–21   Chrysler electronic spark control system. (Chrysler)

vacuum, throttle position, ambient air temperature, and even barometric pressure in some systems. The computer considers all the inputs and then calculates the best amount of spark advance for the engine at that instant. Since the timing is all controlled electronically, there are no centrifugal or vacuum advance mechanisms in the distributor, and no ported vacuum switches or spark delay valves.

This approach has several advantages over the mechanical methods of spark control. First and foremost, it allows much more precise control of ignition timing. Changes can be made almost instantaneously as conditions demand. This keeps emissions to an absolute minimum while enhancing performance and fuel economy. It also eliminates much of the complex plumbing associated with vacuum spark control systems.

Ford has gone through four generations of its Electronic Engine Control (EEC-I, EEC-II, EEC-III, and EEC-IV). Chrysler's system used to be known as Lean Burn (although most mechanics still call it that anyway), but because of some component changes and recalibration, it is now renamed Electronic Spark Control (ESC). General Motor's earliest system was MISAR. It appeared on 1977 and 1978 Olds Toronados. GM then introduced Electronic Spark Selection (ESS) on Cadillacs and Electronic Spark Timing (EST) on Oldsmobiles. In 1980, GM went to its Computer Controlled

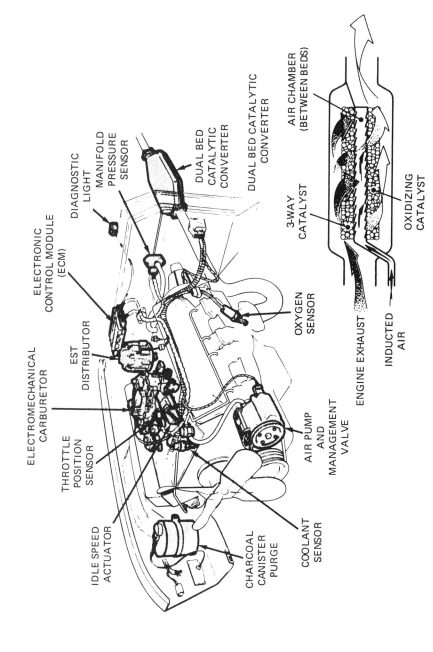

Figure 10–22 General Motors Computer Command Control System. (Chevrolet)

DIAGNOSTIC LIGHT

MANIFOLD PRESSURE SENSOR

ELECTRONIC CONTROL MODULE (ECM)

DUAL BED CATALYTIC CONVERTER

DUAL BED CATALYTIC CONVERTER

ELECTROMECHANICAL CARBURETOR

EST DISTRIBUTOR

OXYGEN SENSOR

THROTTLE POSITION SENSOR

AIR PUMP AND MANAGEMENT VALVE

IDLE SPEED ACTUATOR

CHARCOAL CANISTER PURGE

COOLANT SENSOR

AIR CHAMBER (BETWEEN BEDS)

3-WAY CATALYST

OXIDIZING CATALYST

ENGINE EXHAUST

INDUCTED AIR

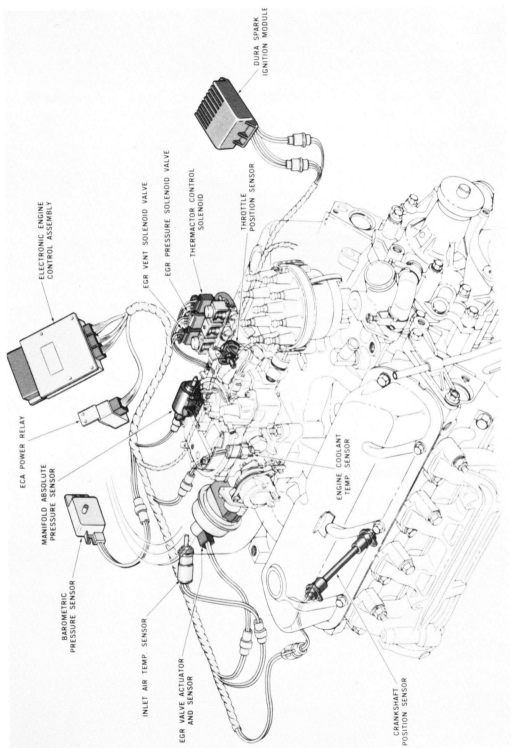

DURA SPARK
IGNITION MODULE

ELECTRONIC ENGINE
CONTROL ASSEMBLY

EGR VENT SOLENOID VALVE

EGR PRESSURE SOLENOID VALVE

THERMACTOR CONTROL
SOLENOID

THROTTLE
POSITION SENSOR

ECA POWER RELAY

MANIFOLD ABSOLUTE
PRESSURE SENSOR

BAROMETRIC
PRESSURE SENSOR

INLET AIR TEMP. SENSOR

EGR VALVE ACTUATOR
AND SENSOR

ENGINE COOLANT
TEMP SENSOR

CRANKSHAFT
POSITION SENSOR

Figure 10–23 Ford Electronic Engine Control system. (Ford Motor Co.)

Catalytic Converter (C-4) system, followed by the Computer Command Control (C-3) in 1981. By the end of 1984, GM had 13 different C-3 systems on its various cars and trucks.

One of the differences between the way the Chrysler system works and Ford's EEC-V and GM's C-3 is that the computer has a built-in clock. When the car is being driven at highway speeds, the computer clock gradually adds spark advance until a maximum is reached several minutes later. The gradual buildup is designed to reduce NO_x formation and improve fuel economy.

Troubleshooting a computerized engine control system is beyond the scope of this book. You need a factory shop manual with the diagnostic procedures, and in some cases special test equipment. Keep in mind, though, the basic principles of spark control. If the spark timing does not advance with increasing rpm (considering any built-in delays), something is wrong with the system. If an engine knocks and pings on acceleration, there is too much advance. Check basic timing, and if it is on the specs, check the EGR valve and air intake system for malfunctions. If everything there is okay and the engine does not have excessive compression due to carbon buildup, there is too much advance. A faulty throttle position sensor, coolant sensor, intake manifold vacuum sensor, or knock sensor may be at fault. Then, again, it might be the computer. To pinpoint the cause on today's complex engine control systems, you must follow the diagnostic procedures in the manual. Unfortunately, there are no shortcuts.

REVIEW

The main points to remember about how ignition affects emissions are:

1. Retarded ignition timing at idle and during deceleration reduces HC in the exhaust.

2. Delayed ignition advance during acceleration reduces NO_x formation.

3. Overadvanced ignition timing can cause detonation which increases HC emissions. Detonation can also damage the engine.

4. How much ignition advance an engine receives and under what circumstances varies tremendously from one application to another. It all depends on the manufacturer's approach to emission control. Generally speaking, though, most engines will have little advance at idle and maximum advance at cruising speed.

11

Other Factors That Influence Emissions

In addition to carburetion and ignition, there are a number of other factors that influence automotive emissions. These include the design and condition of the valve guides and piston rings, the shape of the combustion chamber, the engine's normal operating temperature, camshaft profile, and fuel quality. Changes in any of these factors can cause emissions to increase or decrease depending on the circumstances.

Items such as camshaft profile and the shape of the combustion chamber are determined when the engine is built. How clean or dirty the engine is with respect to these two factors ordinarily does not change over time. Other factors, such as the ability of the rings and valve guides to control oil consumption, are subject to wear. The gradual deterioration of the valve guides and rings will cause emissions to increase over time. Variables such as fuel quality and operating temperature vary constantly, and as a result can cause emissions to change depending on the conditions at any given time. A bad batch of low-octane fuel, for example, can increase HC emissions in the exhaust because of detonation. Hot weather can increase engine operating temperatures, which also increases the tendency to detonate. Let's take a closer look at each of these factors to see how they affect emissions.

VALVE GUIDES AND SEALS

Valve guides are nothing more than a tubular bearing surface through the cylinder head in which the valve stems move up and down. Since this creates a lot of friction, the valve stems must be lubricated by allowing a small amount of oil into the valve guides (Fig. 11-1). The valve guide seals control how much oil dribbles down into

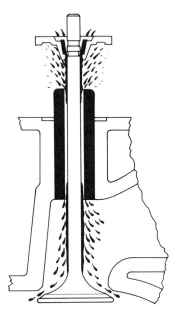

Figure 11-1 Oil leaking past a worn intake valve guide. On every intake stroke, engine vacuum sucks oil down past the guide and into the cylinder. The result is increased HC emissions, high oil consumption and possible oil fouling of the spark plug.

the guides. There are two types of seals: the umbrella or deflector type, which are little circular seals mounted on the valve stem to keep too much oil from splashing or running down into the guides; and the positive seals, which fit around the valve guide boss and scrape excess oil off the valve stems as they move up and down.

When the valve guide seals become worn and/or when the valve guide-to-stem clearances become too great, excessive amounts of oil get past the guides. If the oil gets past an intake valve, it will be sucked into the combustion chamber. If it gets by an exhaust valve, it will be drawn into the exhaust. In either case, HC emissions will be increased. Oil, like gasoline, is a hydrocarbon. But because it is heavier and thicker, it does not vaporize and burn as readily. That is why an engine that uses a lot of oil produces large amounts of HC in the exhaust. If the oil consumption problem is really bad, you will see blue smoke coming out the tailpipe.

One of the myths about controlling oil leakage past the valve guides is that it is a problem only on the intakes. The vacuum in the intake port area during the intake stroke will suck oil through the guides like a straw if the seals are bad. But the same can also happen with the exhaust valves (Fig. 11-2). Even though the exhaust gases are pushed out of the cylinders under pressure, the flow of gases past the exhaust valve guide creates a partial vacuum, which soars right after each exhaust pulse. This can suck oil down the guides and into the exhaust system just as effectively as on the intake side. To make matters worse, exhaust valves usually have larger valve stem-to-guide clearances because they run hotter and need more room for thermal expansion than do intake valves. This can accelerate an oil consumption problem if the exhaust valve seals go bad. Removing the exhaust seals, for example, can cause an engine to use a quart of oil in as little as 450 miles.

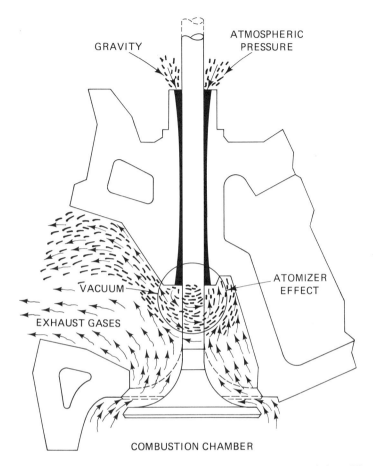

Figure 11–2 Worn exhaust valve guides can also cause excessive HC emissions. When the exhaust exits through the exhaust port, a partial vacuum is created around the bottom of the valve guide. This can suck oil into the exhaust stream to increase HC emissions out the tailpipe.

When an engine is suspected of having worn valve guides and/or seals due to excess HC in the exhaust and high oil consumption, the problem can be cured one of two ways. The cheapest fix is to install new valve guide seals. This can be done without removing the cylinder head. By connecting an air hose to the spark plug hole and filling the cylinder with about 100 psi, the valve will be held shut so that the spring and retainer can be removed. The seal can then be easily replaced. The other alternative is to pull the head and do a complete valve job. Chances are that if the seals are bad, the valves and stems also need attention. When the head is disassembled, the valve stem-to-guide clearances should be carefully checked. If the guides are worn, they will have to be replaced or knurled. They can also be bored out and valves with oversized stems used. In any case, it is an expensive cure for excessive HC emissions.

PISTON RINGS

Like the valve guides and seals, the piston rings also control oil flow and lubrication. The rings limit the amount of oil that lubricates the cylinder walls and ultimately gets into the combustion chambers by scraping the cylinder walls clean with every downstroke, and relubricating them with oil on every upstroke (Fig. 11-3). The top two rings are primarily for compression sealing, while the bottom ring is primarily for oil control. The rings also perform another very important job. They seal the hot combustion gases in the cylinders. Nothing is completely perfect, though, so there is always a small amount of cylinder blowby into the crankcase.

Bad rings can cause a number of problems. If they cannot seal compression adequately, there will be excessive blowby of combustion by-products into the crankcase. These emissions are recycled by the PCV system back into the intake manifold to be reburned in the engine. But if there are more emissions than the engine can handle, the unburned HC will not be burned and will continue on out the tailpipe. Bad rings will also allow oil to enter the cylinders. This will increase oil consumption and raise the amount of HC in the exhaust. If the problem is severe, the car will leave a trail of blue smoke.

Bad rings are diagnosed with a compression check. If the compression readings are low, squirt a few drops of 30-weight oil through the spark plug hole and

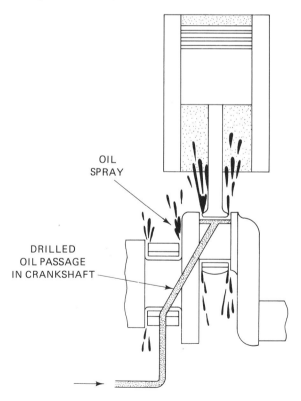

OIL
SPRAY

DRILLED
OIL PASSAGE
IN CRANKSHAFT

Figure 11-3 Excessive oil splash on the cylinder walls due to worn rod bearings can overload the rings and make oil control impossible. If the rings are also worn, oil burning can be severe enough to produce visible blue smoke in the exhaust.

crank the engine over a couple of times. The oil will seal the rings temporarily. Repeat the compression test. If the readings are now much higher, the rings are worn and should be replaced. If there is no change in the compression readings after squirting oil in the cylinders, the valves or head gasket are bad. A cylinder leak-down test can be used to achieve the same results.

The only cure for bad rings is a ring job. The engine must be disassembled so that the rings can be replaced. If the cylinders are found to be out-of-round or to have excessive taper, the block will have to be rebored and oversized pistons installed. Make sure that new rings are installed in accordance with the manufacturer's instructions and that the ring end gap is within specs.

COMBUSTION CHAMBERS

The shape of the combustion chamber plays a role in emissions several ways. When the air/fuel mixture is drawn into the cylinder during the intake stroke, the tiny droplets of gasoline tend to condense on the cylinder walls and combustion chamber surfaces. As the piston travels up on the compression stroke, the droplets of gasoline are scraped off and collect in the pocket just above the top compression ring. When the fuel is ignited, the gasoline trapped in this area around the top piston ring and in the corners of the combustion chamber does not burn. The result is HC in the exhaust.

If a combustion chamber is wedge shaped, that is, it has a large "quench area" where there is very little clearance between the top of the piston and the head (Fig. 11–4a), the flame front will not be able to reach all the corners of the combustion chamber. The result is unburned fuel in the exhaust.

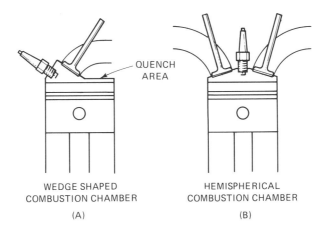

WEDGE SHAPED
COMBUSTION CHAMBER

(A)

HEMISPHERICAL
COMBUSTION CHAMBER

(B)

Figure 11–4 Combustion chamber shape can affect emissions. The chamber (a) is wedge-shaped and has a large quench area. Chamber (b) is hemispherical and has no quench area. The hemi head design typically produces cleaner emissions but is more costly to manufacture.

One of the ways in which engineers can clean up an engine is to redesign the combustion chamber for better efficiency. By relocating the upper compression ring closer to the top of the piston, the pocket above the top ring that tends to collect fuel can be reduced. This helps to lower HC emissions. By reducing the amount of quench area and using a more open combustion chamber design, the flame front can propagate better throughout the cylinder for more complete combustion. Minimizing the surface area of the combustion chamber also reduces fuel condensation on the metal surfaces, and thus helps to lower emissions.

Location of the spark plug hole also plays an important role in how the fuel burns. A centrally located spark plug as in a hemihead design (Fig. 11–4b) is considered best because the flame front can propagate evenly and quickly in all directions once the fuel is ignited. If the spark plug is positioned off to one side, the flame front does not expand uniformly. This can leave little pockets of unburned fuel and increase emissions.

The shape of the combustion chamber can also be designed to induce *swirl* in the air/fuel mixture. This promotes better mixing and more complete combustion. Such a design may require locating the spark plug near the intake valve so that the fuel can be ignited as it swirls past the plug. Swirl is a common feature in many diesel engines.

On Honda's CVCC (Controlled Vortex Combustion Chamber) engine and Mitsubishi's Clean Air Jet system, there is a third valve for each cylinder. The little valve admits fresh air to lead the air/fuel mixture, swirl the charge, and improve flame propagation, all of which results in clean combustion and improved fuel economy.

CAMSHAFT PROFILE

The camshaft controls when the valves open and close. Valve timing, in turn, can affect emissions depending on how the fuel and exhaust flows in and out of the engine. A long-duration cam, for example, with lots of valve overlap will allow unburned fuel to enter the exhaust system. This will increase HC emissions out the tailpipe. Long-duration cams are fine for high-performance applications but not for emissions. Emission cams are therefore typically short-duration cams. In some applications, though, extending the duration of the valve timing can reduce emissions. By holding the exhaust valve open longer, some of the exhaust gases can be drawn back into the cylinder for an EGR effect. The exhaust dilutes the incoming air/fuel mixture to keep temperatures below 2500 °F—the point at which NO_x formation becomes a problem.

The actual valve timing for a given engine will depend on the application, the shape and design of the combustion chambers, and so on. The only real problems a cam can create with respect to emissions is if it develops a flat lobe due to lack of adequate lubrication or incorrect heat treatment of the steel when the cam was manufactured. If a lobe goes flat, the valve that runs off that lobe will not open. The most obvious result will be a severe loss of power in that cylinder. If the flat lobe keeps the intake valve from opening, the cylinder will suck oil past the rings and valve guides and push it out into the exhaust. This will increase HC emissions out the tailpipe. If the lobe affects an exhaust valve, the combustion gases will be forced past the rings into the crankcase. This will increase crankcase emissions and possibly overload the PCV system, again causing an increase in HC out the tailpipe.

FUEL QUALITY AND CONTENT

In terms of quality, a fuel with too low an octane rating for the requirements of a given engine will tend to detonate. This increases HC in the exhaust when the engine is under load and knocking is the loudest. The increase in HC emissions is not great, but the strain detonation places on the engine is a real concern. Severe detonation can crack pistons, burn valves, and cause a lot of expensive headaches, so it is best to correct a detonation problem rather than to ignore it.

The octane requirements of an engine increases with age as carbon deposits accumulate on the pistons and inside the combustion chambers. This raises the effective compression ratio of the engine, and as the compression increases, so do the octane requirements of the engine. If an engine starts to ping and the timing, air/fuel ratio, warm-air-intake system, and EGR valve are not to blame, the odds are that carbon accumulation inside the cylinders is causing detonation. The only cure is to switch to a higher-octane fuel or to knock the deposits loose with a chemical cleaner. If a cleaner does not do the job, the heads may have to be pulled and scraped clean with a wire brush.

In older engines designed to run on high-octane fuel, detonation can be minimized by running a blend of half regular and half unleaded premium. The lead in the regular will raise the octane rating of the unleaded as much as two points. The resulting mixture will have a higher overall octane rating than either the leaded regular or the unleaded premium. The other alternatives are rather expensive. One is to install a water/alcohol vapor injection system. The system sprays a fine mist of water and/or alcohol into the carburetor to boost the octane of the fuel. If this does not work, the only course of action left is to tear the engine apart and install lower-compression pistons, a thicker head gasket, and/or low-compression heads.

As for gasoline content, water in the fuel can cause lean misfiring. And whenever there is misfiring, regardless of cause, unburned gasoline will come out the tailpipe. Alcohol/gasoline fuel blends, typically in the range 10 to 15% and sold under names such as gasohol, regohol, superunleaded with ethanol, and so on, have a slight leaning effect on the air/fuel ratio. In older pre-emission-control engines which typically had rich air/fuel carburetor calibrations, the slight leaning effect may decrease HC and CO emissions somewhat. On late-model engines with lean air/fuel calibration, the leaning effect of the alcohol may or may not change tailpipe emissions.

One gasoline additive that can cause emission problems is lead. Using leaded gasoline in a car designed to run on unleaded fuel (which is illegal) is guaranteed to ruin the catalytic converter. The lead coats the catalyst and renders it useless. It can take only a couple tankfuls of leaded fuel to ruin the converter. After that, the HC and CO will pass right through the converter without being burned. The result is a big increase in HC and CO out the tailpipe. Switching back to unleaded fuel, unfortunately, does not reverse the problem. Once coated with lead, the catalyst is ruined for good. The only cure is to replace the converter on cars with one-piece monolith catalysts or to replace the catalytic pellets in GM converters. Tests have shown that a converter can recover somewhat if only a limited amount of leaded gasoline is burned, say a single tankful. But if leaded gas is used consistently for any length of time, you can kiss the converter goodbye.

Another reason for not using leaded gasoline is that lead itself is a toxic poison. In urban areas where traffic congestion spews large amounts of lead into the atmosphere, lead concentrations can reach levels where lead poisoning becomes a risk. Lead accumulates in plant and animal tissue over time, and can cause a variety of symptoms. The reason lead is used in gasoline is because it boosts the octane rating of the fuel. This allows refiners to use lower grades of gasoline that otherwise would have too low an octane rating for motor fuel.

Leaded gasoline is being phased out by the Environmental Protection Agency. But until there is a total ban on lead (by 1990), it will continue to pose a serious problem.

12

Emissions Testing

The mechanic's primary tool for emissions testing is the exhaust analyzer. Exhaust analysis has come a long way since the days when mechanics depended on Wheatstone-bridge-type combustion analyzers, or air/fuel meters as they were often called. The Wheatstone-type units worked on the principle that different gases have different heat conductivity characteristics. For example, if air is considered to have a thermal conductivity value of 1, hydrogen would be given a value of 7 because it conducts heat seven times faster than air. Carbon dioxide (CO_2), on the other hand, is a poor conductor of heat. It carries heat away about half as fast as air so it would be given a thermal value of $\frac{1}{2}$.

Since the conductivity of a sample of exhaust will vary according to the percentages of different gases it contains, a tester can be calibrated to measure the rates so that a reading of the air/fuel ratio can be determined. The Wheatstone testers compared the amount of heat conducted away from a wire coil exposed to the exhaust to that of another coil exposed to air. The temperature difference between the two coils creates an electrical resistance that causes current to flow through the coil with least resistance. Depending on the direction of the current flow, the meter would either read rich (a high percentage of hydrogen) or lean (a high percentage of carbon dioxide).

The air/fuel analyzers were fine for troubleshooting and adjusting carburetors, but they were not suited to sophisticated emissions analysis. What was needed was a tester that could easily and quickly measure the two key pollutants: unburned hydrocarbons (HC) and carbon monoxide (CO).

INFRARED EXHAUST ANALYZER

The more accurate means of testing exhaust emissions came from the laboratory. The infrared exhaust analyzer is the primary diagnostic tool today for emissions testing (Figs. 12-1 through 12-3), and operates on the principle that different gases absorb different wavelengths of infrared light. Using these values, the analyzer can precisely measure the relative amounts of unburned hydrocarbons and carbon monoxide in the exhaust.

Figure 12-1 Typical infrared exhaust analyzer. Shown is an FMC 905. The analyzer has two scale ranges for HC and CO: 0–400 ppm and 0–2000 ppm for HC and 0–2% and 0–10% CO. (Courtesy FMC Corporation)

The infrared analyzer reads HC and CO contest by splitting a beam of infrared light with a mirror and passing half the beam through a sample of the exhaust and the other half through a reference gas. The beam is chopped on and off by a spinning wheel. The beam then passes through a series of optical filters and strikes a light detector. The detector senses the difference between the two beams and generates an electrical signal that is processed electronically to produce the meter readings. Unburned hydrocarbons are displayed on one meter in parts per million (ppm) and carbon monoxide is displayed on a second meter in percent.

MULTIGAS ANALYSIS

The latest trend in exhaust analyzers is to measure oxygen (O_2) and carbon dioxide (CO_2) as well as HC and/or CO (Figs. 12-4 and 12-5). Being able to read oxygen and/or carbon dioxide in the exhaust is, from a diagnostic standpoint, very helpful on cars that have catalytic converters. Neither oxygen nor carbon dioxide is a pollu-

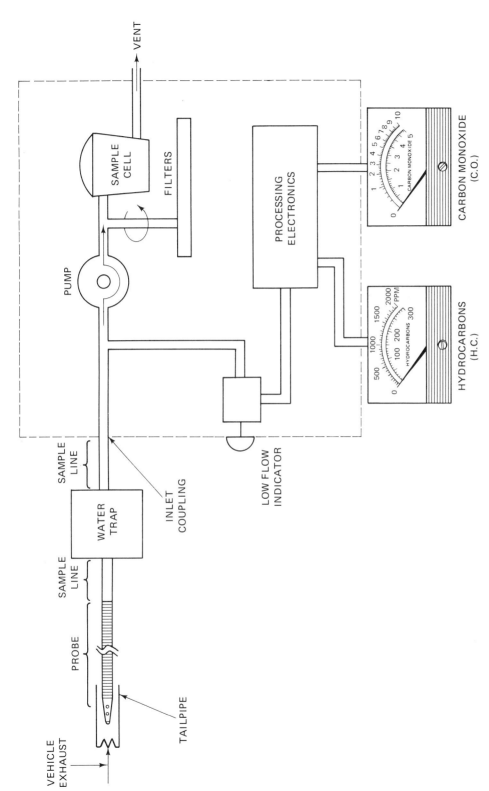

Figure 12–2 Operating schematic of a typical infrared exhaust analyzer. Exhaust gases are drawn from the vehicle's tailpipe into the analyzer where an infrared light beam passes through the gas. The amount of light absorbed depends on the quantity of HC and CO in the exhaust. The results are then displayed on the meters. (Chrysler)

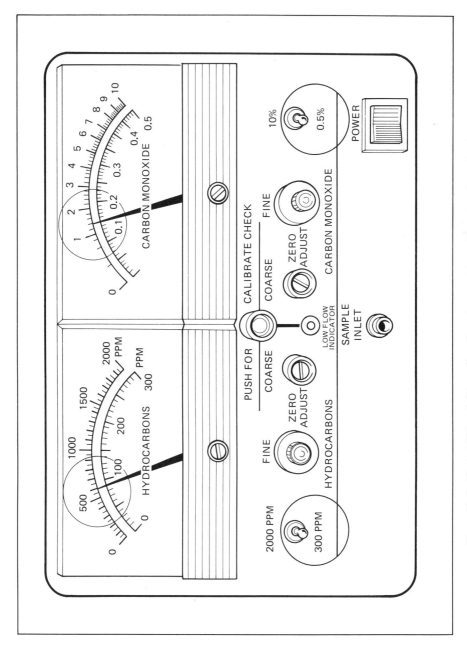

Figure 12-3 Typical HC and CO readings for a late model car. Note that the scale selector switches are set on the lower scales. The reading is 90 ppm HC and 0.15% CO. (Chrysler)

Figure 12-4 A Bear computerized exhaust analyzer. This unit uses a slightly different technology to read the exhaust gases called "non-dispersive" infrared analysis. The results are displayed in digital readouts. (Bear Automotive)

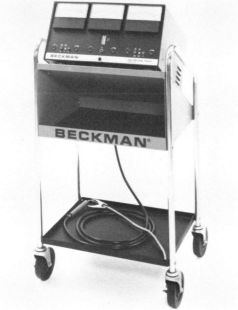

Figure 12-5 A three gas analyzer from Beckman. This unit measures HC, CO, and CO_2. The Model 591 has conventional meter displays. The exhaust pipe probe is coiled up under the unit. (Beckman)

tant, but they are good indicators of combustion efficiency. Using HC and CO alone to measure combustion efficiency is difficult because the converter masks potential problems by converting the HC and CO to water vapor and carbon dioxide. In other words, the converter cleans up the exhaust pollutants so well that some alternative means of measuring combustion efficiency must be used. That is where the oxygen and/or carbon dioxide reading ability comes in (see Fig. 12–6 and Tables 12–1 through 12–3).

The relative proportion of these two gases in the exhaust can tell you whether the air/fuel ratio is right, or whether there is a problem that is causing excessive pollution. As combustion efficiency decreases, the oxygen content will rise and the carbon dioxide content will fall. An engine that is running at a nearly ideal air/fuel ratio of 14.5:1 will show about 14.5 percent carbon dioxide and 2.5 percent oxygen in the

TABLE 12-1 BASIC CO/HC EXHAUST EMISSION DIAGNOSTIC GUIDELINES FOR VEHICLES WITHOUT CATALYTIC CONVERTER[a]

	Speed					
	Idle	1000	2000	Test results	Typical problems	Recommendations
CO				Normal	Defective points/electronic ignition—open	Test using engine analyzer menu
HC	High	High	High	High all speeds	plug wire/failed plug	
CO				Normal	Vacuum leak affecting one cylinder	Check for vacuum leak
HC	High	High	High	Unsteady		
CO	Low	Low	Low	Low	General vacuum leak Should decrease at	Check for general vacuum leak
HC	High	High		High at low rpm	higher rpm	
CO	High			High at idle	PCV valve restricted or carburetor idle mixture	Check the PCV value or adjust the carburetor
HC				Above normal	misadjusted	
CO	High	High	High	High at all rpm	Dirty air cleaner, choke malfunction, or	Replace the air cleaner Service the choke
HC				Above normal	carburetor malfunction	Overhaul the carburetor
CO	High	High		High at low speeds	Rich carburetor adjustment	Adjust the carburetor to specification
HC	High			High at idle		
CO				Normal	Ignition misfire at high speed or floating	Test using engine analyzer menu
HC			High	High at high rpm	exhaust valves	
CO				Normal	Compression loss or vacuum leak	Check compression and manifold vacuum
HC	High			High at idle only		
CO	Low			Very low (idle only)	Lean carburetor adjustment	Adjust carburetor to specification
HC	High			High at idle only		

[a]Engine diagnostic testing should be done with air pump disconnected.

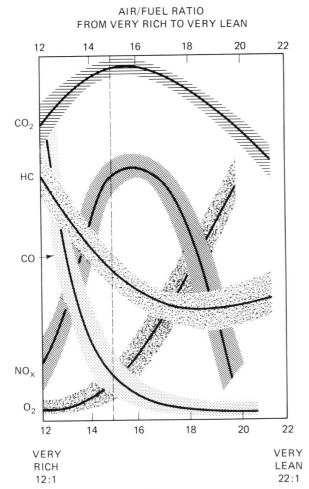

AIR/FUEL RATIO
FROM VERY RICH TO VERY LEAN

Figure 12–6 This chart shows how the composition of the various exhaust gases varies with the air/fuel ratio. The dotted vertical line is at the ideal 14.5:1 ratio. Of these five gases, only HC, CO, and NO_x are pollutants. (Chrysler)

exhaust. Carbon dioxide readings of less than about 13 percent and oxygen readings greater than about 4 or 5 percent indicate poor combustion efficiency. This translates to an over-rich or over-lean air/fuel ratio, poor compression, or an ignition problem.

As for reading HC and CO, keep in mind that a high HC reading means a lot of unburned gasoline and/or oil is passing through the engine. This could be due to a misfiring plug (which can increase the HC reading 10 times over normal), worn piston rings or valve guide seals (which allow oil to be sucked into the combustion chamber), or a super-rich air/fuel mixture.

High carbon monoxide readings mean incomplete combustion, or not enough air to burn the fuel completely. This is due to an over-rich air/fuel mixture. In Chapter 13 we discuss how to troubleshoot emissions problems and what the various readings mean.

TABLE 12-2 INFRARED GAS ANALYZER TESTING.

- Warm engine to operating temperature
- Choke and fast idle cam release
- Insert probe into vehicle's exhaust
- Push RUN button

NOTE: On vehicles equipped with an air pump, the pump output to the exhaust should be stopped for testing.

Test and Conditions	Vehicle Condition	Result	Problems

Idle Speed	Rough Idle	High HC
		Normal CO
			1. Ignition (Misfires)
			  A. Condenser leaking
			  B. Points
			  C. Plugs shorted or fouled
			  D. Plug wires leaking
			     or crossed
			  E. Cap cracked or tracked
			  F. Improper timing
			2. Blowby (Excessive)
			  Do PCV Test

	Rough Idle	High HC
		Low CO
			1. Carburetion (Lean)
			  A. Carburetor mounting
			     tightness
			  B. Mixture setting
			  C. Float operation
			2. Vacuum Leak (Lean)
			  A. Hose connections
			  B. Cracked Hoses
			  C. Manifold leaks
			  D. Gasket leaks

	Rough Idle and
	Black Smoke
	Poor Gas Mileage
		High HC
		High CO
			1. Carburetor (Rich)
			  A. Mixture setting
			  B. Power valve stuck on
			  C. Choke valve
			  D. Float setting
			  E. Air filter restricted
			2. Blowby (Excessive)
			  A. PCV valve sticking
			  B. Worn engine

	Rough Idle	Normal/Low HC
		High CO
			1. Carburetor (Rich)
			  A. Mixture setting
			  B. Choke sticking

1500 to 2000 rpm
Let displays stabilize
	Possible Surging	High HC
		Normal CO
			1. Ignition (Misfire)
			  A. Plug wires leaking
			     or crossed
			  B. Cap cracked or tracked
			  C. Plugs shorted or fouled
			  D. Condenser leaking

136

Test and Conditions	Vehicle Condition	Result	Problems

			2. Timing (Advanced)
			  A. Initial
			  B. Mechanical advance
			  C. Vacuum advance

	Possible Surging	High HC
		Low CO
			1. Carburetion (Lean)
			  A. Float setting
			  B. Jet size
			  C. Mixture setting
			2. Vacuum Leaks (Lean)
			  A. Hose connections
			  B. Manifold leaks
			  C. Carburetor tightness

	Poor Gas Mileage
	Black Smoke
		High HC
		High CO
			1. Carburetion (Rich)
			  A. Float setting
			  B. Jet size
			  C. Power circuit
			  D. Mixture setting
			  E. Choke sticking

Accelerator Pump
  Open and close throttle quickly
  Repeat two or three times
	Hesitation	CO does not increase 1%	1. Accelerator Pump (Lean)
			  A. Linkage
			  B. Pump seal

		CO falls slightly then increases	1. Accelerator Pump (Lean)
			  A. Linkage loose
			  B. Pump seal

Power Valve
  Remove one end of accelerator pump linkage
  Make sure linkage will not bind as throttle is opened
  Increase speed to 1500 to 2000 rpm using step on fast idle cam
  Open throttle quickly and return to same step on fast idle cam
	Poor High Speed Performance	CO does not increase 1% after throttle is opened	1. Power Valve (Lean)
			  A. Sticking
			  B. Spring weak

	Poor Gas Mileage	CO does not return to normal when throttle is closed	1. Power Valve (Rich)
			  A. Sticking
			  B. Vacuum weak or leaking

Vacuum Leaks
  Run engine at a steady speed
  Push choke valve shut or partially block carburetor intake
	Rough Running	HC does not increase	1. Ignition (Misfire)
			  A. Plugs shorted or fouled
			  B. Plug wires leaking or crossed
			  C. Cap tracked
			  D. Points

(Continued on Next Page)

Test and Conditions	Vehicle Condition	Result	Problems
		HC goes lower	1. Vacuum Leak (Lean Misire) 　A. Hose connections 　B. Hose condition 　C. Gasket leaks 　D. Carburetor mounting
PCV System 　Run engine at idle and note CO% 　Pull PCV valve out of valve cover, note CO% 　Block end of valve, note CO%	Oil Leaking	CO unchanged with valve out and unplugged	1. PCV System 　A. PCV hose clogged 　B. PCV valve 　C. PCV carburetor fitting
		HC increase with valve out and unplugged	1. Carburetor (Lean) 　A. Mixture 　B. Float setting
		CO unchanged with valve blocked	1. PCV System 　A. PCV valve 　B. PCV hose 　C. PCV carburetor fitting
EGR System 　Run engine at idle and note CO% 　With a vacuum pump or mechanically open EGR valve slowly watching CO%	Pinging	CO% does not change as EGR valve is opened	1. EGR System 　A. EGR valve diaphragm 　B. EGR stem stuck 　C. EGR passages plugged
AIR System 　Run engine at idle with AIR system working and note HC and CO 　Block air pump output and note HC and CO	Increased Pollution	HC and CO do not increase when output is stopped	1. AIR System 　A. Plumbing to exhaust manifold(s) 　B. Check valve(s) 　C. Pump belt 　D. Pump 　E. Diverter (gulp) valve
Catalytic Converter 　Run engine at idle and note HC	Increased Pollution	HC not below 300 ppm	1. Ignition (Misfire) 　A. Plugs fouled 　B. Plug wires leaking or crossed 　C. Cap tracked or cracked 2. Catalytic Converter 　A. Lead coated 　B. Not functioning
Air Filter 　Run engine at 1500 to 2000 rpm, note HC and CO%	Mileage Loss	CO% decreases more than .5% with element out	1. Filter 　A. Restricted

Test and Conditions	Vehicle Condition	Result	Problems

Remove air cleaner element, note HC and CO%		HC increases with element out	1. Carburetor (Lean)
			   A. Float height
			   B. Mixture setting
			   C. Vacuum leak

Fuel Leaks
  Hold probe ABOVE any suspected leak point
  NOTE: Never let probe draw in fuel as this will damage the analyzer
	Mileage Loss	HC increases	1. Fuel System
			   A. Fuel leak

Exhaust System Leak
  Run engine at idle
  Partially block output of exhaust system
  Move probe around suspected exhaust leaks
  NOTE: To find smaller leaks, partially choke carburetor to increase emissions during test
	Dangerous	HC or CO in any amounts	1. Exhaust System
			   A. Leaks

TABLE 12-3 CONVERTER-EQUIPPED THREE- AND FOUR-GAS ANALYZERS[a]

Gas component	Reference column with air pump on			Test results column with air pump off		
	Speed			Speed		
	Idle	1000	2000	Idle	1000	2000
Correctly adjusted engine						
CO	Low	Low	Low	Low	Low	Low
HC	Low	Low	Low	Low	Low	Low
CO_2	Med	Med	Med	High	High	High
O_2	High	High	High	Low	Low	Low
Rich adjustment of carburetor						
CO	High	High	High	High	High	High
HC	High	High	High	Med	High	High
CO_2	Low	Low	Low	Low	Low	Low
O_2	High	High	High	Low	Low	Low

(Continued on Next Page)

139

TABLE 12-3 CONVERTER-EQUIPPED THREE- AND FOUR-GAS ANALYZERS[a]

Gas component	Reference column with air pump on			Test results column with air pump off		
	Speed			*Speed*		
	Idle	1000	2000	Idle	1000	2000
Lean adjustment of carburetor						
CO	Low	Med	High	Low	Low	Low
HC	High	High	High	High	High	High
CO_2	Low	Low	Low	Low	Low	High
O_2	High	High	Low	High	Med	Low
Vacuum leak (General: PCV hose open)						
CO	Low	Low	Low	Low	Low	Low
HC	High	High	High	High	High	High
CO_2	Low	Low	Low	Low	Low	Med
O_2	High	High	Low	High	High	Low
Vacuum leak (single point)						
CO	Low	Low	Low	Low	Low	Low
HC	High	High	High	High	High	High
CO_2	Low	Low	Low	Low	Low	Low
O_2	High	High	High	High	High	Low
Misfire (bad plug wire)						
CO	Low	Low	Low	Low	Low	Low
HC	High	High	High	High	High	High
CO_2	Low	Low	Low	Low	Low	Low
O_2	High	High	Low	Low	Low	Low

[a]Carbon monoxide: normally ranges between 0 and 0.50%; Hydrocarbons: normally range between 0 and 75 ppm; Oxygen: used to indicate leanness, and normally is less than 7%; Carbon dioxide: used to indicate combustion efficiency, and normally ranges between 7 and 15%.

What about NO_x? Unfortunately, there is no simple and easy way to measure NO_x content in the exhaust. So although NO_x is one of the major pollutants, the only way to determine accurately how much of the stuff there is in the exhaust is in a laboratory. All the mechanic can do is to inspect the EGR valve and related plumbing to make sure that the system is functioning.

BASIC POLLUTION CHECK

When using an infrared exhaust analyzer, whether it is a two-, three-, or four-gas model, there are some basic procedures that should be followed to make sure your readings are accurate.

Before you even touch the meter, inspect the exhaust system on the car you are testing. Check the exhaust manifold, head pipe, converter, muffler, and tailpipe for leaks. Any air leaks, even pinholes, can cause misleading readings on your meter. If you find any problems with the exhaust system, they should be patched or repaired *before* you make your readings.

The engine must also be warmed up and at operating temperature for accurate readings. If the engine is cold, you will probably get abnormally high CO readings because the choke is still partially closed. HC readings may also be higher because the pistons have not swelled up to help seal the combustion chambers.

Okay, if the car is ready, you are ready to take your emissions readings. Although the exact warm-up and calibration procedure will vary somewhat from one analyzer to the next, the following procedure can be used as a general summary. Always follow the analyzer manufacturer's operating instructions for your particular unit.

1. Turn the analyzer on and allow it to warm up for the specified period of time (at least 5 minutes).
2. Zero the meter needles.
3. If equipped with a self-calibration check button, push it to see that the meters are reading as they should. Otherwise, perform the calibration procedure as outlined in your operator's manual.
4. Check the water trap on the probe hose and clean if necessary.
5. Insert the exhaust probe 6 to 12 inches up the tailpipe. Make sure that the probe is well up the tailpipe so that it does not pick up outside air.

   On vehicles with dual exhaust, you should take a reading on both sides to make sure that both banks of cylinders and both catalytic converters are working.

   On some vehicles you can take readings ahead of the converter by opening up a plug in the head pipe or exhaust manifold and taking your readings there.
6. Connect a tachometer to the engine so that you can read idle rpm and high-speed rpm.
7. On cars equipped with air pumps, the air pump must be disabled to prevent air from being pumped into the exhaust manifold. The extra oxygen allows the converter to mask HC and CO readings. The air pump can be disabled by removing the V-belt, by disconnecting and plugging the air supply hose, or by applying vacuum to the diverter valve so that the air is dumped back into the atmosphere. It is possible to get a fairly accurate reading if you can take your measurements before the catalytic converter starts cooking. But once it warms up, it will clean up the HC and CO unless the air pump is disconnected.
8. Look at the emissions label under the hood to determine the correct idle speed and timing for the vehicle being tested.
9. Take your readings. Allow about 10 seconds between readings for the meters

to stabilize. Excessive idle time can load up the exhaust with excessive carbon. To compensate, accelerate the engine to about 2000 to 2500 rpm and hold there for 5 to 10 seconds. This will blow out the system and allow accurate readings. Allow at least 30 seconds after a blow-out procedure for the meters to stabilize.

IDLE TEST

Most emissions specs are for engine idle speeds since this is when the engine is the dirtiest. As rpm increases, the HC and CO levels should decrease.

If the engine idle speed and timing are correct, the HC and CO readings should agree with the emissions label under the hood or with the specifications supplied in your testing manual. See Table 12–4 for a guide to acceptable emissions levels. Note that these levels are a guide only, as specific vehicles may have higher or lower legal limits. Each state is also free to set its own emissions standards for inspection purposes, and these standards may be tougher or more lenient than these.

If your emissions check reveals a high HC and/or CO reading, you will have to check out the cause of the problem (see Chapter 13). The first item to check is idle speed and basic timing. If either of these is off, it could affect the tailpipe readings. Readjust to the manufacturer's specs and repeat the emissions check. If the readings are still too high, the engine has a problem that requires more detailed diagnosis.

If you are using a meter that also reads oxygen and/or carbon dioxide, refer to Table 12–5 to interpret combustion efficiency.

TABLE 12-4 BASIC CO/HC EXHAUST
EMISSION DIAGNOSTIC GUIDELINES FOR
VEHICLES WITHOUT A CATALYTIC
CONVERTERS.[a]

Use this chart with your exhaust analyzer to pinpoint emissions problems.

| Vehicle year | Normal limits at idle | |
	HC (ppm)	CO (%)
1967 and earlier	300–500	2.5–3.0
1968–1969	200–300	2.0–2.5
1970–1972	150–250	1.5–2.0
1973–1974	100–200	1.0–1.5
1975–1978[b]	50–100	0.5–1.0
1979–1983[b]	50	0.0–0.5

[a]These figures are a guide only and should not be used in place of the specs on the emissions decal or testing manual.
[b]Readings with air pump disconnected.

TABLE 12-5 FOUR GAS DIAGNOSTIC CHART
FOR VEHICLES EQUIPPED WITH CATALYTIC
CONVERTER.[a]

Air/fuel ratio	Percent CO	Percent CO_2
16.0:1 (lean)	0.0	13.5
15.0:1	0.02	14.3
14.5:1 (ideal)	0.12	14.4
13.8:1	1.8	13.5
13.4:1	2.9	13.0
12.8:1	4.5	12.1
12.3:1	5.8	11.9
11.0:1	9.4	9.0
10:5:1 (rich)	10.0	8.0

[a]Multigas analyzers that read carbon dioxide (CO_2)
can be used to calculate the air/fuel ratio.

HIGH-SPEED TEST

An emissions check at 2500 rpm when performed in conjunction with an idle test
can help pinpoint problems. Both HC and CO readings should be lower at 2500 rpm
than at idle. If the readings are the same or higher, a problem exists. A higher CO
reading indicates too high a carb float level, a fuel-saturated float, a stuck power
valve, oversize carb jets, excessive fuel pump pressure, a leaky needle valve or seat,
a dirty air filter, or an obstruction in the air-intake system. A high HC reading indicates a high-speed miss due to a weak ignition coil, faulty plugs or wires, incorrect
timing or dwell, or a lean misfire due to a vacuum leak or over-lean air/fuel mixture.
See Chapter 13 for troubleshooting emissions problems.

CARBURETOR TESTS

An exhaust analyzer can also be used to test both carburetor power valve and accelerator pump operation. Although a complete test of the power valve can be made
only with the engine under load (preferably on a dynamometer), you can check to
see if the valve is functioning at all under no-load conditions. With the transmission
in neutral or park, disconnect the accelerator pump linkage and adjust the idle speed
to approximately 1500 rpm. Observe the CO reading, then snap the throttle fully
open and return it to fast idle. The CO reading should increase by nearly 2 percent
when the engine is throttled, but it should then return to the fast idle reading. If no
change in CO is observed under these conditions, the power valve is not functioning.

To test the operation of the accelerator pump, idle the engine in park or neutral
until the CO reading stabilizes. Snap the throttle open a small amount. This should
cause an increase in CO of 1 to 2 percent or more. The accelerator pump is bad if

the CO level drops off sharply before starting to rise. Check for excessive play in the accelerator pump, the cup to see that it is seating, and the check valves to see if they are leaking.

PCV SYSTEM TEST

Your analyzer can also be used to check the operation of the positive crankcase ventilation system. Run the engine at idle and make sure that it is at operating temperature. Note the normal CO reading on the meter. Pull the PCV valve from the valve cover and allow it to hang free. Note the CO reading. It should drop below the CO level at idle. If it drops more than five percentage points, it indicates excessive fuel in the crankcase. The fuel can come from a variety of sources. A leaky mechanical fuel pump on some engines can leak fuel into the crankcase. A misfiring plug and/or worn rings will allow raw fuel to blow by the rings into the crankcase. A leaky intake manifold gasket can allow fuel into the engine. Or excessive cranking during cold weather can flood the engine with raw fuel, some of which will find its way past the rings and into the crankcase. Whatever the source, gasoline vapors will be drawn through the PCV system and into the intake manifold. This will cause a rich air/fuel ratio, which in turn can cause higher than normal CO readings at the tailpipe.

If you find an engine with severe fuel contamination in the crankcase, it is wise to change the engine oil after the source of the problem has been fixed. Fuel-diluted oil makes a lousy lubricant. A lot of short-trip stop-and-go driving can cause fuel vapors to accumulate in the crankcase because the engine never gets warm enough to evaporate the fumes completely. Running the car at highway speeds for 15 to 30 minutes is usually sufficient to dry out the oil.

To test the PCV system for blockage, put your thumb over the open end of the PCV valve (Fig. 12–7). You should feel vacuum and you should note an increase in the CO reading. Plugging the end of the valve with your finger shuts off the flow of air into the intake manifold and causes a richer air/fuel ratio. If there is no change and you feel no vacuum, you have a plugged PCV valve or hose. Carbon accumulation under the carburetor inside the intake manifold can also sometimes block off the PCV connection.

EXHAUST SYSTEM CHECK

The analyzer can be used to detect exhaust leaks in the exhaust system by moving the probe along the length of the system while the engine is running. If there are any leaks, the fumes will be drawn up the probe and into the analyzer. If there are no leaks, the HC and CO meters should stay on zero.

You can also check the passenger compartment and trunk for dangerous carbon monoxide accumulations with the analyzer. Allow the engine to idle for 15 to 20 minutes with the car windows and trunk closed. Crack the window and insert the

Figure 12-7   Temporarily blocking the PCV valve should cause the CO reading to increase.

probe into the passenger compartment, plugging the rest of the crack with a shop towel. Do the same for the trunk. If there is any leakage of carbon monoxide into the car, you will get a reading on the CO meter.

PROPANE-ENRICHED IDLE CHECK

The idle mixture adjustment on late-model carburetors with limiter caps (Fig. 12–8) or sealed idle mixture screws can be checked using the propane enrichment procedure. Propane is fed into the carburetor through a vacuum connection or the air cleaner. This enriches the air/fuel mixture and causes the idle speed to increase—up to a point where the mixture becomes too rich and the speed drops.

The enriched rpm specification on the underhood emissions label tells you at what rpm the idle should peak out if the idle mixture screws are adjusted properly. If the idle mixture is too lean, the rpm drop will occur at a lower speed than the specs. If the mixture is too rich, the speed drop will happen at a higher rpm. Your analyzer will show a higher-than-normal CO reading if the idle mixture is set too rich. If the idle mixture is too lean, CO will be less, but HC may increase if the mixture is lean enough to cause lean misfire. If hand choking the carb at idle lowers a high HC reading, the mixture is too lean.

COMBUSTION LEAK TEST

If you have an engine that is overheating due to a mysterious loss of coolant, you can use your analyzer to detect combustion leaks into the cooling system. Remove the radiator cap and allow the engine to idle. Place the analyzer probe above the

IDLE MIXTURE SCREW

Figure 12-8 Idle mixture adjustment is critical to low emissions at idle. On late model engines with idle mixture limiter caps or sealed screws, the mixture can be checked with the propane enrichment procedure. Too rich increases CO. Too lean decreases CO but can increase HC.

coolant level in the radiator opening. Do not stick the probe into the coolant. Rev the engine up a couple of times or put it under load for a few seconds by placing it in gear with the brakes on (automatic transmission only). Combustion leaks through the head, head gasket, block, or intake manifold gaskets will show up on the HC meter as the tiny bubbles of exhaust gas find their way through the cooling system.

13

Troubleshooting Emissions

Problems

Troubleshooting emissions problems is not that difficult if you have a good understanding of the basic principles of emissions cause and effect. In other words, if you know that excessive hydrocarbon (HC) emissions are caused by anything that allows unburned gasoline to pass through the engine or causes ignition misfiring or missing, and that excessive carbon monoxide (CO) emissions are due to anything that causes an over-rich air/fuel ratio, you can usually isolate the source of the problem by a simple process of elimination. You start with the most likely causes and work down the list from there.

To help speed your diagnosis, you should make use of every available diagnostic tool at your disposal. An ignition analyzer or oscilloscope is a tremendous timesaver when it comes to pinpointing ignition problems. Many of the new "smart" diagnostic computers will not only tell you that an engine has an emissions problem but will also do the diagnosis for you and give you a list of items to check. A unit such as a Sun 1215 or Allen Smartscope (Fig. 13-1) can pick up such things as a bad spark plug, low coil voltage, low cylinder compression, lean or rich air/fuel mixture, timing problem, and so on. Lacking such sophisticated equipment is a handicap in this day and age, but you can still achieve the same results with simple hand-held test equipment such as a vacuum gauge, compression gauge, and tach/dwell meter.

Of course, the one piece of equipment that is absolutely essential to emissions troubleshooting is your exhaust analyzer. Without it, you have no way of making certain that you have correctly isolated and repaired the source of the problem. In many instances an emissions problem may be due to more than one cause. For example, an engine pumping excessive HC into the tailpipe may have a couple of fouled spark plugs—but it might also have worn valve guide seals, piston rings, and low

Figure 13-1 Using a computerized engine analyzer can be a real time saver when troubleshooting emissions problems.

coil voltage. If you fix only the most obvious problem, the spark plugs, the amount of HC in the exhaust will be reduced but probably not enough to pass inspection. Because of that, it is important that you nail down *all* the causes of an emission problem—not just the most obvious one or ones.

THE ART OF ANALYSIS

Emissions testing and troubleshooting has always been surrounded by a certain amount of controversy because some people say that it is more of an art than a science. Sometimes the same car will show different emissions levels when tested on different exhaust analyzers. The reason for this is that there is always a certain amount of variance from one equipment manufacturer to another when it comes to calibrating the analyzers. All exhaust analyzers are supposed to read the same, but in the real world there is always a small amount of variance between analyzers—even on units off the same assembly line. What is more, analyzers, like any other piece of calibrated equipment, can get out of adjustment. That is why it is extremely important to keep your analyzer properly maintained and adjusted at all times.

Problems can also result when one facility does the initial emissions testing and another does the repairs. In states such as Arizona and others, state-run centralized inspection stations do the emissions testing. If a car is rejected for excessive emissions, the motorist is free to take the vehicle to the garage of his or her choice to have it repaired. But the owner must then take it back to the state inspection station for retesting. If the garage's analyzer is not calibrated the same as the state's, the garage may tell the motorist there is nothing wrong with the car or may misdiagnose the problem. In either case, the motorist will receive another rejection slip when he or she takes the car back for retesting.

In other states, such as New York, Massachusetts, California, and Pennsylvania, a decentralized program for inspection is used. Repair shops and service stations are licensed by the state to do the inspections, and each must purchase an "approved" exhaust analyzer. The analyzers are connected to a computerized data recording system so that any variance from the norm can be quickly detected. The analyzers must also be maintained according to a strict schedule. The advantage of this approach is that the facility that does the testing is usually the facility that does the fixing. So repairs can be checked and verified quickly to make sure that emissions problems have indeed been corrected.

But in spite of our best efforts to make emissions testing a science rather than an art, factors do creep in that can cause misleading results. Weather, driving conditions, and even the quality or type of fuel used can create emissions problems in engines that otherwise would pass an inspection.

For instance, if a car spends a lot of time idling while waiting in line at an inspection station, it is possible for a buildup of carbon to cause excessive HC readings. The same car, if driven right in off the highway, might pass the test.

On some late-model cars with feedback electronic engine controls, excessive idling can cause the oxygen sensor to cool off. This causes the engine control system to switch to the open-loop mode of operation and a rich air/fuel mixture. If this happens when the car is tested, it will show high CO emissions. Running the engine at fast idle will warm the oxygen sensor up enough that it will produce a signal and cause the system to switch back to closed loop. If everything is in good working condition, the CO readings should drop back to normal.

On hot summer days, warm air is less dense than usual. This can cause a slight increase in the richness of the air/fuel ratio (less air means a richer mixture). If the engine is a borderline case to begin with, the added strain of hot weather might be enough to push it over the brink as far as emissions are concerned. The same engine would produce lower CO readings on a cooler day.

The type of fuel used can also affect emissions. Gasohol, a blend of 90 percent gasoline and 10 pecent ethanol alcohol, typically has a leaning effect on carburetion because alcohol contains about one-third less heat energy per gallon as gasoline. For ordinary driving, there is usually no detectable difference in performance. But when it comes to emissions, there can be a noticeable difference—sometimes better, sometimes worse. It all depends on how the engine is calibrated. In older vehicles with richer air/fuel ratios, the use of gasohol typically lowers CO emissions. But in some late-model cars with extremely lean air/fuel ratios, the use of gasohol can lean the mixture to the point where occasional lean misfire can occur. This would increase HC emissions.

As you can see, emissions testing can be a mixed can of worms. There will always be a certain amount of variance between analyzers and even on the same vehicle from one day to the next. But these variances should be small. Emissions troubleshooting, therefore, may involve making a judgment call as to whether or not a borderline case has a real problem. Most of the time, though, an emissions problem is obvious enough to produce a significant and repeatable reading on your or anybody else's analyzer.

PRIMARY CAUSES OF HIGH HC EMISSIONS

When troubleshooting high HC emissions, keep in mind that HC means unburned gasoline in the exhaust. Any condition in the engine that allows unburned gasoline to pass through the combustion chamber or interferes with ignition will increase HC (see Fig. 13-2). A fouled spark plug or leaking exhaust valve, for example, will increase HC tenfold. An occasional plug misfire or a mild case of oil consumption, on the other hand, may increase HC only slightly.

Most engines suffering from extremely high amounts of HC in the exhaust will probably have a noticeable miss—due to fouled spark plug or plugs, bad plug wires, cracked distributor cap, bad exhaust valve, or blown head gasket. An ignition problem will show up on your oscilloscope. A high firing voltage for any given cylinder could mean that the plug is not firing because of excessive resistance in the plug wire or plug. A corroded rotor or distributor cap can also cause high firing voltage. A low-voltage reading on a cylinder could mean that the spark is being shorted somewhere. A compression problem due to a bad valve or head gasket can be isolated by performing a cylinder power balance test or by doing a compression test.

The following are the most common causes of high HC emissions (you can also refer to Tables 12-1, 12-2 and 12-3 in the preceding chapter):

1. *Fouled spark plug(s).* No spark equals no combustion (see Fig. 13-3).
2. *Worn spark plugs or incorrect gap.* Too wide a gap requires more voltage than the ignition system can supply. The result is no spark and no combustion. If the gap is too narrow, lean mixtures of air and fuel will not be ignited.
3. *Ignition misfiring.* Misfiring is due to cracked distributor cap, rotor arcing, low coil voltage, incorrect ignition dwell, high resistance in plug wires, shorted or burned plug wires, poor plug wire terminal connections, or intermittent short in ignition module.
4. *Oil burning.* This results from worn piston rings, scored cylinder walls, cracked pistons or rings, worn valve guides, or worn valve seals (see Fig. 13-4).
5. *Compression leak into exhaust manifold.* This occurs because of pitted, bent, or cracked exhaust valve; broken exhaust valve spring; weak exhaust valve spring; or insufficient exhaust valve clearance.
6. *Compression loss.* This is caused by a blown head gasket.
7. *Excessively rich air/fuel ratio.* Engine is flooded with fuel it cannot burn due to choke stuck shut, carb fuel leak, carb flooding, or extreme restriction in air intake or air filter.

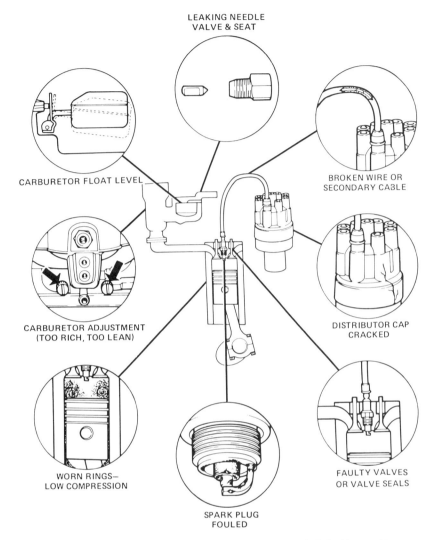

Figure 13-2 The primary causes of high HC emissions include ignition problems, poor compression, oil consumption, and excessively rich fuel mixtures. (Chrysler)

8. *Extremely lean air/fuel ratio.* This causes lean misfiring (insufficient fuel for combustion). Problem is due to vacuum leaks, fuel starved carburetor, or low float level in carb. A very lean air/fuel mixture requires more voltage to fire, so the problem may also be traced to low plug voltage.

When troubleshooting a high-HC-emissions problem, check the ignition system first, then the compression, to rule out the more obvious possibilities. "Reading" the spark plugs is one way to tell whether or not the air/fuel mixture is overly rich.

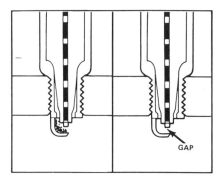

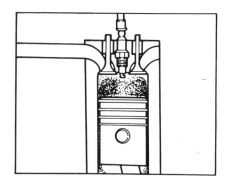

Figure 13-3 The plug on the left is fouled with deposits while the one on the right is worn to the point where the gap is too wide for the spark to jump. Either condition will cause increased HC emissions because the mixture isn't ignited. (Chrysler)

Figure 13-4 Oil consumption past worn rings or valve guide seals is another common source of high HC emissions. Worn or damaged exhaust valves can also allow compression loss into the exhaust system. (Chrysler)

A sooty, black coating on the plugs indicates an overrich mixture (Fig. 13–5). Conversely, a white-to-yellow glazed appearance of the plug tip insulators indicates overly lean. Normal is light brown to tan deposits. Looking at the tailpipe can also give you a clue to the engine's carburetion. A rich mixture will cause black sooty deposits, with soot possibly visible in the exhaust when the engine is revved up.

If you notice black sooty deposits in the tailpipe of a car equipped with a catalytic converter, it means that the converter is not doing its job—possibly because it has been fouled with leaded gasoline or because it has been fouled or burned out because of an excessively rich fuel mixture. A misfiring plug, for example, will overload the converter with fuel. This can produce one of two results: Either the converter clogs up with carbon or it burns out because of the high temperatures created when it tries to burn up the additional HC. In either case, the only cure is to replace the converter and repair the source of the rich-fuel condition. Replacing just the con-

Figure 13-5 A carbon fouled spark plug. Note the heavy black sooty deposits on the plug tip and insulator resulting from an excessively rich fuel mixture. (Champion Spark Plug Co.)

verter will not solve the problem because it, too, will quickly succumb to the over-rich fuel condition. As for lead poisoning, examine the gas tank filler cap to see if someone has punched out the unleaded restrictor plate. It takes only a couple of tankfuls of leaded gasoline to ruin the converter—and the only cure is to replace the converter and stop using leaded fuel.

PRIMARY CAUSES OF HIGH CO EMISSIONS

High CO readings mean that the air/fuel ratio is too rich. There is not enough oxygen available to burn the fuel completely, so some of the carbon will be converted to carbon monoxide instead of carbon dioxide (see Fig. 13-6). That extra atom of oxygen per molecule makes the difference between a deadly poison and a harmless gas.

When troubleshooting the cause of high CO emissions, look for anything that might cause a rich air/fuel mixture: problems in the carburetor or fuel injection system that allow more fuel than usual to enter the engine, or restrictions in the PCV or AIR emissions control systems.

The most likely causes of high CO emissions are:

1. *Carburetor idle mixture adjusted too rich.* Use the propane enrichment idle speed drop test to confirm the condition before fiddling with sealed idle mixture adjustments. It will save you a lot of time and trouble—especially if the high CO readings are due to something else.

2. *Idle speed below specs.* Check the idle speed and repeat the idle test if necessary. Readjust as required to bring the rpm reading up to specs.

3. *Malfunctioning or misadjusted choke.* On a warm engine that has been idling for at least several minutes, the choke should be fully open. If not, check the mechanism for binding. Clean with carb cleaner as required to free the mechanism. On older cars with adjustable choke housings, check to see that the choke is adjusted to specs. Other problems that can cause overchoking of the engine include restricted or disconnected heat riser tube from the exhaust manifold to the choke housing; a broken bimetallic spring inside the choke housing; a shorted heating element on electrically heated chokes or a dead wire to the heating element; or dirt or corrosion in the choke vacuum mechanism causing binding (see Fig. 13-7).

4. *Malfunctioning heated air intake system.* The heated air system is designed to provide warm air to the carb during engine warm-up and cold-weather operation to aid fuel vaporization. But if it sticks in the position that draws up warm air from around the exhaust manifold, the air/fuel mixture will be richened because warm air is less dense than cool air. Check the air control flap inside the snorkel for proper operation, and check the vacuum connections, vacuum motor, and thermostatic control. If the ambient air temperature is over 80 °F and the engine is warmed up, the door should be open to outside air.

5. *Restricted air intake.* This may be due to a dirty air filter (Fig. 13-8) or clogged

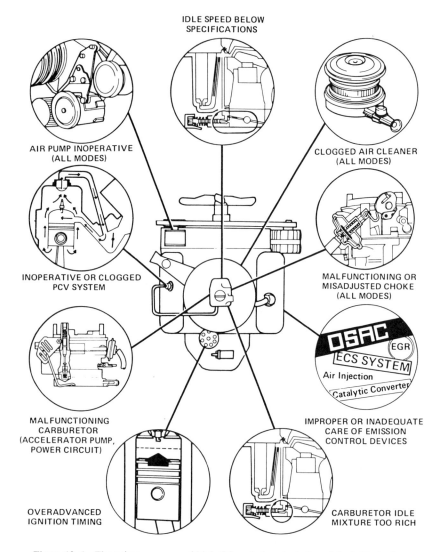

Figure 13-6   The primary causes of high CO emissions include a rich air/fuel mixture, overadvanced ignition timing and an inoperative AIR system. (Chrysler)

air intake system. Leaves or debris in the air intake can block airflow and cause high CO readings. Such an engine will also exhibit lack of high-speed power.

6. *Rich air/fuel ratio due to carburetor malfunction or misadjustment.* Check for such things as leaky needle valve and seat, high fuel level in bowl due to leaky float or fuel-saturated float; excessive fuel pump pressure; internal carb leaks in metering circuits or gaskets (Fig. 13-9); leaky power valve; too-large main jets for carb application; or dirt-clogged air passages in metering circuit.

Figure 13-7 Choke problems are probably one of the most common causes of high CO idle readings.

Figure 13-8 A dirt clogged air filter can restrict air flow and upset the air/fuel ratio. The result is higher CO emissions.

On late-model cars with feedback engine control systems, a variety of problems can cause the carburetor to run rich. On General Motors C-4 and C-3 systems, a fuel mixture control solenoid inside the carburetor controls the air/fuel ratio. An electrical signal from the computer tells the carb how rich or lean to make the mixture, depending on what the oxygen sensor in the exhaust manifold tells the computer. The longer the dwell signal, the leaner the mixture. The shorter the dwell, the richer the mixture. If the oxygen sensor is not warm enough to produce a signal (engine still cold or excessive idling), the system goes into open-loop operation and the carb runs rich. If the oxygen sensor is defective (due to wear or contamination from using leaded gasoline or carbon fouling), the system will run in open loop and be too rich. If the coolant sensor is defective and keeps telling the computer that the engine is

Figure 13-9   Another common cause of high CO emissions is fuel saturated plastic carburetor floats. As the float becomes heavier, it sinks lower than the bowl. This allows more fuel into the carb and causes the mixture to run rich. If the float is heavier than a new one, replace it.

cold, the system will stay in open loop. If there is a fault in the wiring between the computer and mixture control solenoid, the carb will run too rich because there is not sufficient dwell signal to make it run lean. A defective computer can also give the carb the wrong signal. To diagnose one of these systems, you will need to use the fault codes and diagnostic trees in the factory shop manual.

On late-model Ford products with feedback engine controls, a slightly different approach is used. Instead of using an electrically controlled solenoid in the carburetor to control the air/fuel mixture, Ford uses a vacuum-controlled metering rod. No vacuum signal to the metering rod equals full rich. Five inches or more vacuum equals the lean position. The air/fuel mixture is controlled by varying the vacuum signal through an electrically controlled solenoid on the fender or firewall. As with the GM system, a problem with the oxygen sensor, coolant sensor, computer, or control solenoid can cause an over-rich mixture. Again, you will have to refer to the factory diagnostic procedure to pinpoint the faulty component.

On vehicles with either throttle body electronic fuel injection or multipoint electronic fuel injection, a leaky fuel injector can cause a rich air/fuel ratio. With throttle body injection, all engine cylinders are affected. But on multipoint systems, only the cylinder supplied by the faulty injector runs rich.

A leaking injector in a multipoint fuel injection system can be found one of several ways. One way is to inspect the spark plugs. The plug from the cylinder with the leaking injector will show signs of carbon or soot fouling. Another diagnostic technique is to perform a power balance test. The engine is run at idle while each cylinder is momentarily shorted out by removing a spark plug wire (on computer systems with idle speed control, the idle speed control motor must be unplugged before the test; otherwise, the computer will attempt to compensate for the shorted cylinder by increasing idle speed). The rpm drop for each cylinder is noted, and the cylinder or cylinders that produce either the least or most change in rpm should be checked further. If a cylinder produces little or no change in rpm, it means that it is not contributing much to the engine's power output. This could be due to a fouled spark plug (too much fuel from the leaky injector), a bad valve, or an inoperative injector. In some cases, an injector that was dumping excessive quantities of fuel into the cylinder would cause an rpm drop. But if an injector is running slightly rich, the cylinder actually puts out *more* power than the others. This is why you must look for either an increase or decrease in rpm rate during the power balance test.

Another type of power balance test can be done to isolate a faulty fuel injector also. Here each injector electrical lead is momentarily disconnected while the engine is idling. Disconnecting the lead should cut off all fuel to the cylinder if the injector is not leaking, and by watching the exhaust HC emissions with an infrared exhaust analyzer, you can note if there is a difference in the drop of HC emissions between cylinders. The one that drops the least or increases is the one with the leaking injector. This test works best on four-cylinder engines where each cylinder contributes a significant portion to the overall emissions of the engine. For best results, the exhaust sample should be taken ahead of the catalytic converter so that the converter cannot "mask" the readings.

1. *Plugged PCV valve or hose.* Any restriction in the PCV system will reduce the amount of air being drawn into the intake manifold. This has the same effect as increasing the richness of the air/fuel ratio. If you pull the PCV valve from the valve cover and place your finger over the end, you should feel a vacuum, and at the same time you should notice a decrease in idle rpm and note an increase in CO on your exhaust analyzer meter. The PCV valve itself should rattle if you shake it (Fig. 13–10). If none of the items just mentioned applies to the engine you are examining, you have found a problem. It is also a good idea to check the PCV filter inside the air cleaner if so equipped, as a restriction here can cause the same type of problem.

2. *Inoperative AIR system.* This results from a slipping V-belt on air pump, defective air pump, loose or reversed AIR plumbing connections, faulty diverter valve, or diverter valve control (Fig. 13–11). The extra oxygen that the AIR system pumps into the exhaust system helps convert the carbon monoxide to carbon dioxide. If this supply of air is blocked, CO readings will increase. You can check the AIR system by temporarily disconnecting the plumbing from the back of the pump to see that air is coming out of the pump, or by applying vacuum to the diverter valve (or disconnecting vacuum as the case may be) to see if air dumps into the atmosphere. If air is coming out of the pump and is passing

Figure 13-10 The wrong PCV valve for a given application can also cause emissions problems by allowing either too much or not enough air flow. This upsets the idle air/fuel mixture and can cause increased CO emissions (rich mixture due to insufficient air flow through valve) or increased HC emissions (lean misfire due to too much air flow through the PCV valve).

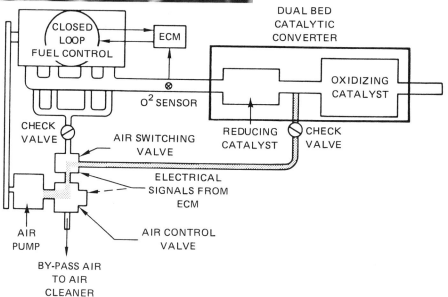

Figure 13-11 Late model General Motors three-way converter system for reducing NO_x, HC, and CO. A problem in the air control diverter valve can foul up the air supply to the catalysts, increasing NO_x, HC, or CO. (Chevrolet)

through the diverter valve as it should, the only other possibility might be blockage in the air lines due to carbon or damage.

On PULSE air systems (those that use no pumps but use vacuum pulses in the exhaust system to suck fresh oxygen into the exhaust manifold), the only parts that can cause a problem are the PULSE air valve and the muffler. If the intake muffler becomes clogged, no air will be drawn into the system. Same applies to the PULSE air valve. If it jams shut, air will not get through. If the system is functioning, you should hear what sounds like a motorboat with the top of the air cleaner removed.

3. *Overadvanced ignition timing.* Too much advance at idle can cause the CO readings to increase. Check the timing and adjust as necessary to bring it within specs. As with high HC emissions, a fouled catalytic converter will allow excessive CO to pass through the exhaust system unscathed.

4. *Fuel in the crankcase.* This occurs because of short-trip stop-and-go driving or a fuel leak through a fuel pump. Fuel will be sucked into the intake system through the PCV valve. This will cause a rich mixture.

CAUSES OF HIGH HC AND CO EMISSIONS

If you find an engine that pumps out high levels of both HC and CO, suspect a major problem in the carburetor, AIR system, or converter. If the engine is overloaded with fuel, it will not burn completely and some of it may pass right on through untouched. If there is not sufficient extra oxygen waiting for the combustion gases in the exhaust system and if the converter cannot get the job done of converting the HC and CO to carbon dioxide and water vapor, you will have an emissions problem.

Of course, you may have an engine that suffers from a number of problems. A super-rich air/fuel mixture will quickly foul the plugs. This, in turn, will aggravate the emissions problem by adding HC to the CO. Then, too, you might have an old engine that has an oil consumption problem (high HC) that is a little too rich on the fuel (high CO). Add to that the fact that it has not been tuned for who knows how long (fouled plugs, timing off, dwell off, idle speed, and mix off), and you can see you might have a whole list of items that are contributing to the emissions problem.

DETECTING AIR AND VACUUM LEAKS

An air or vacuum leak into the carburetor, intake manifold, or anywhere in the vacuum plumbing will cause the air/fuel mixture to run lean. This can cause drivability problems such as hesitation, stalling, poor fuel economy, and whistling noises. It can also make carburetor adjustments extremely difficult unless the problem is isolated and corrected.

If you find that the CO meter readings do not respond when adjusting the idle mixture screw on the carburetor, outside air is probably being drawn into the system through a vacuum leak. You can use your analyzer to verify this condition by removing the air cleaner and manually choking the engine by pushing the choke almost shut with your hand while the engine is idling. If the engine speed increases and the CO level rises while you choke the engine with the palm of your hand, an air leak is indicated. If the HC level rises at the same time, the carburetor is probably adjusted much too lean. If the HC level rises sharply and the engine labors and threatens to stall, excessive blowby is indicated. For this, you will have to do a compression check or a cylinder leakage test.

You can isolate a vacuum leak by removing one vacuum hose at a time and applying vacuum with a hand pump. If the vacuum holds (such as to the distributor vacuum advance diaphragm), the connection is good. Repeat the test until you find the leaky hose or component. Connecting a vacuum gauge to the far end of each vacuum line can also be used to tell whether or not vacuum is holding within the line.

Another way to find a vacuum leak is to disconnect and plug one vacuum hose at a time while watching the HC/CO meter. If the CO level rises when a hose is disconnected and plugged, you have found the problem.

TROUBLESHOOTING NO_x EMISSIONS

Although the only way to measure NO_x emissions accurately is in the laboratory, you can troubleshoot the EGR system, which is the primary emission control system for NO_x emissions. If the system is working correctly, the EGR valve should remain closed when the engine is cold and during idle. It should open only after the engine has reached operating temperature and under part-throttle conditions. When ported vacuum is supplied to the EGR valve, you should be able to see the valve stem rise, indicating that the valve has opened up. On many late-model cars, the EGR valve has a built-in back-pressure sensor which causes the valve to cycle up and down. This is normal for this type of valve.

If the valve fails to open or cycle at part-throttle operation, check to see that vacuum is reaching the valve. If it is, try applying vacuum to the EGR valve directly to see if it opens. If it does not open, the diaphragm is probably bad or the valve corroded. In either case, you will have to replace it.

Often an EGR valve will stick open, or not close completely because of carbon accumulation around the valve seat. This will cause rough idling and stalling. Incorrect routing of the vacuum hose to the EGR valve can also cause this same condition if the hose is connected directly to manifold vacuum. This would cause the valve to open at idle, which is exactly what you do not want because the engine cannot handle the added flow of exhaust gases at slow speed.

On late-model cars with three-way catalytic converters, a problem with the primary converter or its air supply can increase NO_x emissions (see Fig. 13-9). If the diverter valve on the AIR system is not working correctly, airflow to the primary catalyst may not be timed correctly to reduce NO_x (the diverter valve on such cars is controlled by the computer). Use of leaded gasoline will also ruin the effectiveness of the catalyst. Since you cannot measure NO_x emissions directly with your exhaust analyzer, all you can do is inspect the engine to see that the EGR and AIR systems are intact and appear to be working, and that someone has not pumped leaded gasoline into the fuel tank.

TROUBLESHOOTING SUMMARY

One of the aftermarket trade magazines for professional mechanics ran some tests using two-gas and four-gas exhaust analyzers to see if the additional carbon dioxide and oxygen readings were of any help in finding engine problems, and to see if a two-gas analyzer alone could provide enough useful information on late-model cars. Two vehicles were selected for the tests, a 1974 Pinto (no converter) and a 1981 Ford Escort (converter equipped).

The Pinto tests provided good diagnostic information using just the two-gas analyzer. At idle the readings were 0.29 percent CO and 180 ppm HC. With a vacuum leak introduced, the readings were 1150 ppm HC and 0.15 percent CO. This was exactly what one would expect—low CO due to the introduction of more oxygen, and high HC due to the resulting lean misfire. With one plug wire removed, the HC reading jumped to 1920 ppm and the CO remained at 0.29 percent—again exactly as expected. The HC emissions went up because one cylinder quit firing but CO remained the same because the air/fuel ratio was not changed.

At 2500 rpm, the Pinto showed HC content of 95 ppm and a CO reading of 0.42 percent. An increase in CO at 2500 rpm in comparison to the idle reading usually suggests that the carburetor power valve is leaking or that the main metering jets are too large. And this would probably be the diagnosis given using a two-gas analyzer. But when the car was tested using a four-gas analyzer, a different diagnosis emerged.

The idle readings were 6 percent oxygen and 10.9 percent carbon dioxide. These readings indicate rather inefficient combustion (a properly tuned engine should be around 2.5 percent oxygen and 14.5 percent carbon dioxide). When compared to the 0.29 percent CO idle reading, which is quite low for what a 1974 Pinto is supposed to show, the results suggest an overly lean idle mixture. Sure enough, as the Pinto continued to idle during the test, it started to develop a lean misfire, causing the HC readings to jump to 650 ppm. The oxygen reading increased to 8.5 percent.

At 2500 rpm, the Pinto produced 4 percent oxygen and 13.1 percent carbon dioxide, which was considered quite efficient for this engine. When compared to the HC and CO figures, the results show that the fuel mixture at 2500 rpm is near optimum. The fact that the CO reading was higher than at idle reconfirms the conclusion that the idle mix was too lean rather than the 2500 rpm mix being too rich, as originally believed.

When the 1981 Escort was tested, it produced 40 ppm HC and 0.01 percent CO at idle. The testers found that they could produce significant changes in the HC or CO readings in the tailpipe by introducing malfunctions in the engine before the converter warmed up. But once the converter warmed up and started doing its job, the changes in HC and CO resulting from disconnected plug wires or vacuum leaks was negligible. For example, immediately after starting the car, one plug wire was pulled. This caused the HC reading to jump to 250 ppm. But within a few minutes, the converter began to work it down and managed to clean it up to 45 ppm.

The air pump had to be disconnected to get valid oxygen and carbon dioxide readings when the four-gas analyzer was used. When the plug wire was disconnected, the four-gas analyzer picked it up. The oxygen reading went from 5.5 percent at idle up to 8 percent, and the carbon dioxide reading fell from 12.5 percent to 10.5 percent. The introduction of a vacuum leak did little to affect the HC and CO readings, but caused a substantial change in the oxygen and carbon dioxide readings. The vacuum leak also caused more of a change in the oxygen reading than carbon dioxide reading. A 2500-rpm check on the Escort for excellent combustion efficiency with an oxygen reading of 3 percent and carbon dioxide of 14.3 percent.

So what does all this mean? It means one of the best ways to learn to diagnose emissions problems is to play around with your analyzer. Hook it up to a car and "create" some emissions problems. Pull a plug wire, change the timing, change the idle mix, and so on, and see what effect it has on the HC, CO, O_2, and CO_2 readings. Experience is the best teacher.

14

Troubleshooting Drivability Problems That Are Emissions Related

The emission control systems that are used on all cars sold in this country have done a wonderful job of reducing the amount of pollution our society spews into the atmosphere. But in engineering, just as in economics, there is no free lunch. The advances the automakers have made toward the elimination of HC, CO, and NO_x from motor vehicles were expensive in terms of money, but even most costly from the standpoint of complication and the potential for drivability and performance problems that are difficult to diagnose.

PAST SIMPLICITY

When you open the hood of a pre-control-era car, everything looks relatively simple and straightfoward. Usually, there is only one vacuum hose that is related to engine performance. It runs between the carburetor and the distributor vacuum advance unit. With an automatic choke, there is in some cases another short hose that routes manifold vacuum to the choke pull-off diaphragm. There are no spark delay valves, dual-diaphragm distributors, or thermostatic vacuum switches, and there is not even a hint of the complex EGR arrangements we have today. Nor are there air pumps, charcoal canisters, PCV plumbing, or any suggestion of electronics whatsoever.

This simplicity makes diagnosis and repair amazingly simple relative to the situation today. Drivability problems could usually be traced very quickly to an ignition breakdown, a faulty choke, or some internal carburetor malfunction. Beyond that, a mechanic would suspect that the trouble resided inside the engine itself in the form of a burned valve, a bad timing chain, or similar problem.

MODERN COMPLICATIONS

Now, of course, things are entirely different. All those add-on components and the plumbing that connects them require a great deal of study and experience to be truly understood. There are many more places to look for the cause of a problem, and if mechanics do not know the theory and operation of the various emissions systems present on a car, they have to resort to trial and error for diagnosis, which does not make them popular among their patrons. Or, they may not be able to find the cause of a problem at all, and that will surely diminish the customers' confidence.

Although it would be impossible in a single book to explain specifically every conceivable emissions-related drivability problem on the multitude of different systems that are on our highways, we can give you enough general theory and specific practical instances to allow you to extrapolate to the particular car you have in front of you.

A high percentage of drivability problems are fuel-system related. That is because a modern carburetor supplies the engine with a much leaner blend of air and gasoline than was used in pre-emission-control days, and a mixture that is too lean will misfire. This means that today's power plants are on the ragged edge of a burnable ratio, and anything that leans the blend out the least bit more than it should be will cause missing, hesitation, and/or stalling.

A very concrete example of this can be found in a problem that many mechanics have been confronted with: a dead miss at idle in the number 6 cylinder of a Chrysler slant six-cylinder engine. Ignition timing, spark, the condition of the distributor cap, the spark plug wires, and the plug itself all check out perfectly, so compression becomes suspect. A test is taken and it is found that cylinder 6 is the equal of the others. So why is that cylinder alone missing?

The answer is a surprise to most mechanics: The slant six has separate intake manifold runners for each of its cylinders. The one that goes to cylinder 6 has a vacuum takeoff fitting screwed into it. If just one of the several vacuum lines that are connected to this takeoff has even a small crack in it, it will admit enough extra air into the cylinder to make the mixture too lean to burn, which proves how closely calibrated carburetion is today.

So, especially with that maze of hoses that jam a modern engine compartment, always look for a vacuum leak when you run into a car that will not idle properly.

HESITATION

Since the automakers started battling the twin problems of emissions and fuel inefficiency, the number of complaints about hesitation has grown tremendously. Everything has got to be just right or these lean mixtures will not fire dependably.

The first thing to do when you encounter a conveyance that stumbles or bogs on acceleration is to open the hood and take a cursory look around. Is the heated-air duct intact (Fig. 14–1)? Are any vacuum lines cracked or disconnected? Is the heat riser valve frozen in the open position? If the engine is cold, start it up and see that

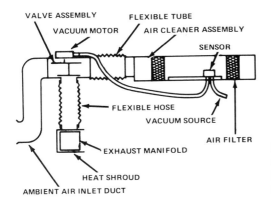

VALVE ASSEMBLY   FLEXIBLE TUBE
VACUUM MOTOR   AIR CLEANER ASSEMBLY
SENSOR
FLEXIBLE HOSE
VACUUM SOURCE
AIR FILTER
EXHAUST MANIFOLD
HEAT SHROUD
AMBIENT AIR INLET DUCT

Figure 14-1 A faulty heated-air-intake system is a common cause of hesitation. It can also cause stalling and affect idle quality. (American Motors Corp.)

the snorkel door closes (a wet rag placed on the air bleed valve will simulate chilly conditions). Next, remove the air cleaner and make sure that the vacuum hoses which operate the heated-air-intake setup are in good shape and properly connected. We have fixed numerous, supposedly incurable cases of hesitation just by attending to this system.

The choke deserves special attention. Obviously, it should be closed when the engine is cold and open when it is warm, but there is more to it. Check the setting of the vacuum pull-off and the adjustment of the fast idle screw.

Since it is very easy for an engine to fail a government emission certification test in the first 30 seconds of the cycle because of the choke, it is calibrated to open so rapidly that the enigne's hold on life is very tenuous for a while after startup. Therefore, many "cold hesitation" complaints can be traced to the choke.

With the power plant shut off, peer into the carburetor while you open and close the throttle. If you do not see a robust stream of gas, you can be pretty sure that the accelerator pump is the culprit.

POWER CIRCUIT

A carburetor's power circuit is designed to increase the amount of fuel the engine gets as vacuum drops, or, with mechanical systems, as the throttle plates move toward wide open. Any malfunction here will definitely cause a flat spot.

Make sure that any metering rod or accelerator pump linkage adjustment is correct, and that the float is set high enough.

TIMING

Using a timing light to see if the initial spark lead is set correctly, then look for the proper amount of advance as the engine is speeded up. If you are dubious about the amount the marks move, apply vacuum to the distributor advance unit and see what happens. If there is no change under the light, a new advance unit is in order.

With most transmission or speed-controlled spark systems, there is a provision for allowing vacuum to reach the distributor when the engine is cold regardless of what gear is engaged or how fast the car is rolling. To check this out, plug a vacuum gauge into the line, start the engine cold, and rev it. You should get a reading until normal operating temperature is reached. Then, raise the car safely, put the transmission in Drive or High and see that you get vacuum when accelerating from cruising velocity.

EGR systems are often designed so that they do not interfere with drivability while the engine is cold. Using the gauge again, see that there is no vacuum to the valve until the power plant is warmed up when the rpm is held at about 2000.

STALLING

In a way, stalling is terminal hesitation, so all the things mentioned above should be considered. The most important question to ask yourself here is: When does it stall?

If it is right after startup with the engine cold, the most obvious area of investigation is the choke and fast idle setting. There is a good chance that the thermostatic coil is adjusted improperly or has lost its ability to move the choke plate far enough. Then there is the choke pull-off—maybe it is set so that it opens the choke plate too far when the engine is fired up. Check the specifications and adjust it as necessary. Next, set the fast idle to specs making sure that you have got the screw on the proper step of the cam.

If it stalls in cold weather regardless of how long the engine has been running, the heated-air-intake system may be at fault. If it supplies only unheated air, it could be so dense that it leans out the mixture too much to permit the engine to keep running when the throttle is opened suddenly.

An EGR valve that gets a vacuum signal even when the power plant is cold could be the cause of stalling. Using a vacuum gauge at the EGR valve hose, see that there is no reading until the power plant has warmed up. If you get a reading with the engine cold, check out the thermostatic valve or switch in the system.

An engine that stalls when the throttle is opened, then allowed to snap closed, may have a defective dashpot on its carburetor (Fig. 14–2).

Especially with today's higher operating temperatures, vapor lock is a distinct possibility, so see if the fuel line has been rerouted too near an exhaust manifold or other hot component. If everything appears to be in its original position, why is the engine getting so hot that it makes gasoline boil in the lines? Perhaps the coolant is low, the fan clutch is defective, or on front-wheel-drive cars with electric radiator fans, the thermostatic switch never closes to complete the circuit (this is a common problem). Retarded ignition timing can also cause an engine to run hot.

Beyond that, think about restricted fuel supply, an intermittent electrical problem in the ignition system, and if the trouble occurs mostly on damp days, deteriorated secondary ignition components.

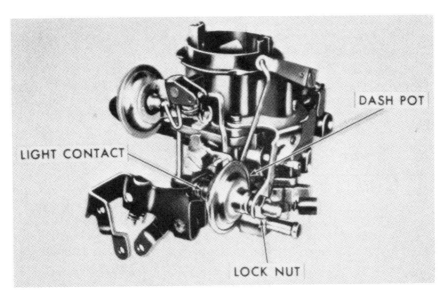

Figure 14–2 An improperly adjusted dashpot can cause stalling. (Chevrolet)

ROUGH IDLING

A poor idle is very annoying, and the cause is sometimes hard to trace. To begin, establish a baseline by setting everything to specifications. While the timing light is connected, take special note of how much the timing marks waver. Some Chrysler engines, for example, have nylon distributor drive gears, and when these wear out, timing varies, which upsets idle quality.

The traditional practice of pulling and reinstalling one plug wire at a time to see if one cylinder does not affect idle is still useful. But with catalytic-converter-equipped engines, it must be done cautiously and quickly, or all that unburned mixture may overheat the catalyst. Remove each wire for as short a time as possible, then reconnect it and run the engine at fast idle for 10 seconds before moving to the next cylinder.

If you have zeroed in on a particular cylinder, remove the spark plug and examine it and its cable. Then take a compression test. If the reading is low, check all the other cylinders to find out if a valve and/or ring job is necessary.

Believe it or not, the heated-air-intake system can cause rough idling. If it is always in the heat-off position so that it allows only cold, dense air to reach the carburetor, it can lean out the blend enough to affect smoothness. That is one of the reasons it is necessary to check idle and adjust mixture screws with the air cleaner installed.

A faulty idle stop solenoid may be allowing the engine to idle too slowly, but you will find that out when you check all the settings against specifications.

A PCV valve that is jammed in the wide-open position can change the way a car idles. A vacuum leak, as mentioned above, is certainly a good possibility, and your vacuum gauge will help you find it.

If the EGR valve sticks open, idle will certainly be affected. If it is open wide, the engine may not idle at all. Check this possibility by seeing if the EGR valve plunger moves freely, and perhaps by pushing its stem toward the closed direction.

HARD TO START

Hard starting can usually be attributed to poor compression or a fuel or ignition problem. If the engine is difficult to fire up when it is cold, the choke is the first thing to check. If the trouble appears only when the weather is damp, the secondary ignition components should be suspect. If it happens when the engine is hot, first find out if it is too hot, then look for anything that might cause flooding, and for maladjusted ignition timing. Of emissions controls, the two most likely to cause this problem are an open EGR valve or a defect in the electronic engine control system if the car is so equipped.

Backfiring on deceleration in cars with air injection can usually be traced to the diverter or other valve that controls this system. It is probably allowing air pump output to reach the exhaust system even under high-vacuum conditions so that the rich decel mixture gets enough oxygen to explode.

DIESELING

Modern engines run hotter than those of the past, and usually their throttle plates are open more at idle to allow a lot of air in to keep the mixture lean (the timing may be set relatively late to keep idle speed low enough). These two factors, combined with a propensity for self-ignition due to low-octane gas, contribute to a most unsatisfactory syndrome—*dieseling*—the term used to describe what happens when you shut off the ignition but the engine keeps running, very roughly. It is amazing how long this can continue, and it does not do the engine a bit of good.

Idle speed is the first thing to check. If there is an idle stop solenoid, see that the lower setting is indeed low enough. We have seen many cases where people have adjusted the idle speed with the throttle screw instead of the solenoid, which eliminated the benefits the device provides. By allowing the throttle plates to close far more than they would at normal idle speed, the solenoid cuts down the flow of the one thing the engine needs to run—air—so it dies with dignity.

A switch to better fuel may help chronic dieselers, but the most important thing is to shut it off as slowly as you can while it is running. That means in gear with an automatic, and with your foot off the accelerator for several seconds.

PINGSVILLE

Detonation, also known as *spark knock* and *pinging,* is an extremely prevalent problem today, with automakers trying to get every mile possible out of a gallon of gasoline and oil companies giving us lower-octane fuel than they did in the past when they could add lead.

Initial spark timing and the advance curve are the first things to check. Often, you will find that the spark lead has been advanced several degrees by someone who was trying to improve performance and fuel mileage.

Another very common cause of detonation is an inoperative EGR system (Fig. 14–3). Since the recirculation of exhaust gases dilutes the air/fuel charge and so lowers peak combustion temperatures, it is a great help in reducing spark knock (which is caused by high temperatures and pressures).

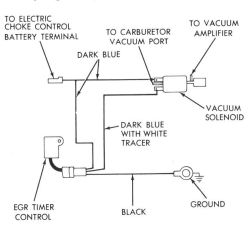

TO ELECTRIC CHOKE CONTROL BATTERY TERMINAL

TO CARBURETOR VACUUM PORT

TO VACUUM AMPLIFIER

DARK BLUE

VACUUM SOLENOID

DARK BLUE WITH WHITE TRACER

EGR TIMER CONTROL

BLACK

GROUND

Figure 14–3 Anything that might keep the EGR valve from opening, such as the time delay circuit, can contribute to detonation. (Chrysler)

An engine that runs too hot will be prone to detonate, so make sure that the cooling system is in good order. A frequent reason for overheating in vehicles with electric radiator fans is a bad thermoswitch, but this is frequently overlooked. Substituting a thermostat with a lower temperature rating than the stock unit (say, a 180 °F unit instead of a 195 °F) can ease pinging, but it may have other detrimental effects.

A heated-air-inlet system that is stuck in the heat-on position can contribute to the problem by raising the temperature of the incoming air, which will bring the charge that much closer to the detonation point.

Deposits that form in the combustion chamber take up space, so they raise the compression ratio and promote pinging. Although fuel additives can reduce these deposits in cars that use regular gas, they are much less useful against the kind of stuff that accumulates with unleaded gas.

Light, occasional pinging will not cause any damage and shows that the engine is running as close to peak efficiency as possible. If it is heavy and constant, however, it can give the engine a severe beating. So when you encounter a case of chronic detonation and everything under the hood is in good order, it might be wise to switch to unleaded premium gas, even though it is more expensive.

BRAIN PROBLEMS

In many cars that are equipped with electronic engine controls, such as Ford's EEC, GM's C-3, or Chrysler's Lean Burn, a defect in the sensors or the computer can cause many of the drivability complaints described above. Unfortunately, the checkout procedures are tailored so specifically to each system, and there are so many systems, that we would have to write an entire book on just that subject—and it would become obsolete every year. So all we can recommend is that you get the right service data for the car in question. We can, however, give you one general rule: Usually, the problem is not in the expensive electronic components, but elsewhere, so be very certain that you have not missed anything before you replace any sophisticated hardware.

15

Tuning-Up Emission-Controlled Engines

In the days before antipollution equipment appeared on cars, a tune-up was basically a matter of replacing the spark plugs, points, and maybe the condenser, then setting the ignition timing, the idle speed and mixture, and if the car had one, adjusting the automatic choke and fast idle. The secondary ignition system—rotor, distributor cap, and spark plug cables—was thoroughly inspected if the mechanic was conscientious. If the power plant had mechanical, nonhydraulic valve lifters, the valve lash might also be set to specifications. Replacement of the air filter element, or a thorough cleaning of the mesh in oil-bath-type air filters, and servicing the fuel filter were frequently included.

DIFFERENT

With the advent of emission controls, that routine started to change. Some items were added to the list, and some were deleted. On pre-1975 domestics and many later imports, ignition service comprised the same work that it always had. But with the appearance of catalytic converters and breakerless electronic ignition in 1975, it was no longer necessary routinely to replace anything but spark plugs, and timing could be set once at the factory and never needed to be adjusted again because there was no point wear to make it vary.

Limiter caps on the idle mixture screws that reduced the amount of adjustment allowed appeared quite early in the fight against air pollution (Fig. 15–1). Then came plugged or locked mixture adjustments that were pretty well tamperproof (Fig. 15–2), and closed-loop feedback carburetors and fuel injection are now becoming common.

FULL-RICH POSITION
COUNTERCLOCKWISE

FULL-LEAN POSITION
CLOCKWISE

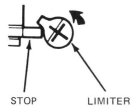

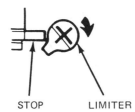

STOP   LIMITER   STOP   LIMITER

Figure 15-1 Limiter caps on the idle mixture screws usually allow only about 270° of adjustment. (Ford)

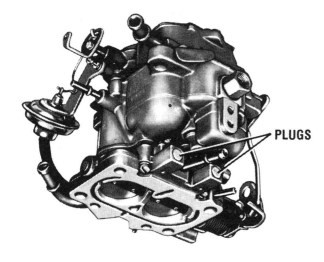

PLUGS

Figure 15-2 Later models have tamper-proof idle mixture screws. The carburetor must be removed to extract the plugs and change the mixture setting, which is not necessary in ordinary service. (Chrysler)

But all that does not necessarily mean that a thorough tune-up now requires much less work. Inspecting the emission control hardware and plumbing and perhaps running functional tests is always a good idea—and one that requires a great deal of new knowledge. Anyone can just replace spark plugs (although there is a right way and a wrong way to do it), but we believe that the definition of a tune-up should include more than that, especially since drivability complaints are much more common on tightly controlled engines than on those of the past, which were designed to run smoothly and the amount of pollution be damned.

PRIMARY

We begin with the system that produces the spark. On pre-electronic-ignition cars (1974 and older domestics; most imports used points until later years), replacement

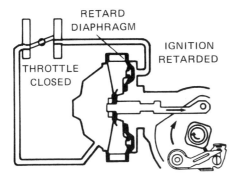

Figure 15-3 Many modern distributors use an advance/retard vacuum unit. At idle and closed throttle decel, manifold vacuum retards timing. (Volkswagen of America)

of the points and condenser is basically the same as it has been for decades. It is timing that became more critical. Many distributors incorporated vacuum units that both advanced and retarded the spark. For example, a common design has separate chambers on either side of a diaphragm that is linked to the plate on which the points are mounted (Fig. 15-3). The outer chamber is connected to a carburetor port above the throttle plates so that it advances the timing when the throttle is opened, as in most ordinary configurations. But the inner chamber is hooked up to a source of manifold vacuum so that it retards the spark at idle and during deceleration in order to help control emissions (similar configurations have been used on some later electronic ignition distributors).

This makes it absolutely necessary that you know for sure if the initial timing setting should be done with the advance or retard hoses connected or disconnected on the particular model at hand. The specifications label in the engine compartment will give you this information. Also, some makes that have only a single-action vacuum unit (Honda, for instance) should have their timing adjusted with the hose connected.

STABILITY

The main reason the automakers adopted electronic ignition was not to produce a stronger spark (in fact, some systems, Chrysler's for example, produce relatively low voltage), but to take advantage of the fact that timing never varies. In a point system, spark is produced when the points open, so timing is affected by the dwell angle or gap width. The longer the dwell or narrower the gap, the later the spark. So, as the rubbing block on the movable point wears, timing becomes retarded (Fig. 15-4). Obviously, without points this situation is eliminated (Fig. 15-5).

Of course, just because timing does not change over time with electronic ignition is no guarantee that it was adjusted properly in the first place, so it is necessary to at least check this setting with a timing light when performing a tune-up (Fig. 15-6).

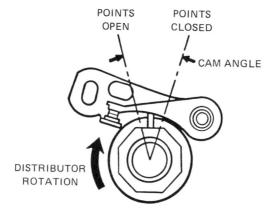

POINTS
OPEN

POINTS
CLOSED

CAM ANGLE

DISTRIBUTOR
ROTATION

Figure 15-4 As the movable point's rubbing block wears, dwell increases retarding timing. (Chrysler)

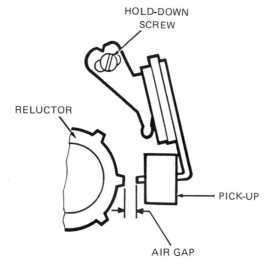

HOLD-DOWN
SCREW

RELUCTOR

PICK-UP

AIR GAP

Figure 15-5 But with electronic ignition, there is no wear—primary circuit control is by means of the varying magnetic field between the reluctor and the pickup. (Chrysler)

ELECTRONIC ADVANCE

Because an engine's spark advance curve is so critical to the amount of all three pollutants that it pumps out, and due to the limitations of mechanical and vacuum devices, automotive engineers have created various types of electronic advance systems. These make use of electronic logic and various sensors to tailor the spark much more accurately to conditions than could be done with mechanical/vacuum setups. Chrysler's Electronic Lean Burn (ELB) was the first example of this, and it allowed engines to use less exhaust gas recirculation (EGR) and still achieve lower levels of NO_x emissions (Fig. 15-7). The other makers followed suit shortly thereafter, so many cars on the road today have electronic spark advance control.

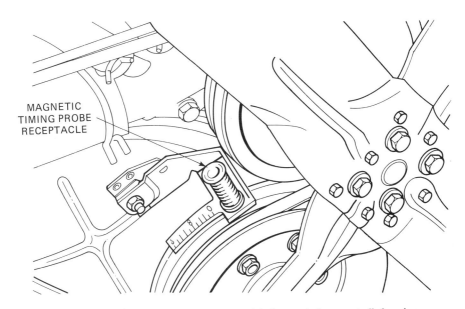

Figure 15-6 Ignition timing is especially critical on emissions controlled engines. Many are equipped with receptacles for use with electronic timing meters that give a direct numerical read-out of spark timing. (Chrysler)

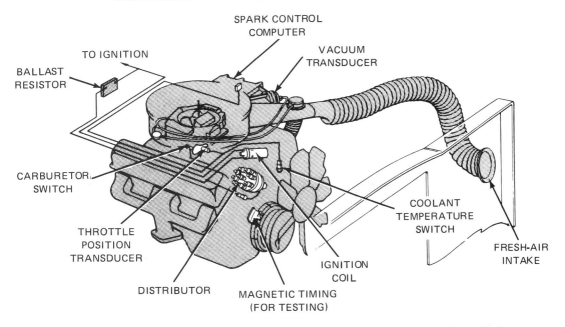

Figure 15-7 Electronic spark control sets ignition timing during all modes of operation by means of sensors and a computer. (Chrysler)

It is beyond the scope of this book to provide you with the checkout procedures for all these systems. Suffice it to say that if there is a performance or fuel mileage problem, the proper factory service troubleshooting information should be consulted before expensive electronic components are replaced. Often, the problem is due to nothing more than a bad electrical connection or a cracked vacuum line.

HIGH-VOLTAGE SIDE

The secondary ignition circuit comprises the rotor, distributor cap, spark plug and coil wires, and, of course, half of the coil. On pre-electronic-ignition cars with emission controls, service of these components is basically the same as it always was, except that the higher temperatures found under the hood relative to uncontrolled models can cause the wires to deteriorate faster.

But with most electronic ignition systems, things are different. The higher voltage produced is naturally that much more prone to escaping. In other words, the strong electrical pressure will "leak" out at any opportunity. Contaminated or cracked insulation, accumulations of dirt on the cap (Fig. 15–8), and/or dampness can all contribute to this situation. Certainly, secondary ignition components have been improved to cope with this extra burden—GM's 8-mm silicone wires, for example—but the potential for problems is still there.

To illustrate: We encountered a 1974 model car that had points but was equipped with an aftermarket capacitive discharge ignition system that produced voltage on a par with that of, say, GM's HEI (high-energy ignition). The engine would absolutely not start in wet weather, but as soon as the CD was taken out of the circuit, the power plant fired up and ran perfectly. It was a textbook case of the secondary components not being capable of handling the voltage the system generated.

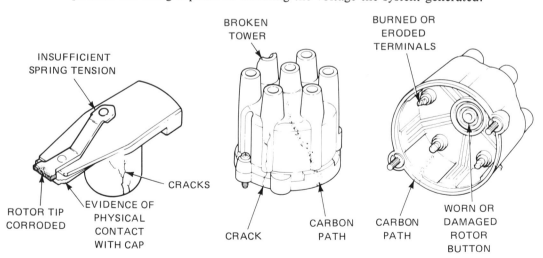

Figure 15–8   The distributor cap and rotor deserve attention at every tune-up, especially with high-voltage ignition systems. (American Motors Corp.)

PLUGS

Spark plugs are the business end of the secondary system. Until the advent of electronic ignition, plug service was basically the same as it had always been: a matter of periodic replacement (at least every 12,000 miles), or occasional cleaning, filing, and regapping. Designs and heat ranges were altered to accommodate emission-controlled engines, but this did not affect procedures or maintenance intervals.

With electronic ignition and unleaded gas, however, the life expectancy of a typical spark plug rose dramatically. Deposits on the insulator nose, which act as shunts that bleed voltage away from the gap, were much less likely to form. This, combined with the fast rise time of the voltage surge with electronic primary circuit switching, allowed plugs to keep firing dependably for a much longer time. Now some automakers are actually recommending plug replacement only every 60,000 miles.

The wide-gap plug (up to 0.080 inch) was another development. This helped assure that the lean antipollution mixture in the cylinder was indeed ignited. Never, by the way, attempt to set ordinary plugs to wide-gap specs, because the electrodes will not be square and thus will erode unevenly.

In the days when cars were permitted to spew pollutants into the atmosphere without limit, the insulator nose of a plug from a cylinder that was firing properly was supposed to appear light brown. As mixtures became leaner, the color went to tan, and now it should be almost white.

A NEW LEASE

A word about cleaning spark plugs is appropriate here. The main goal of the process is to eliminate the deposits that can act like shunts on the insulator nose. Solvent and a brush or sandblasting are the usual methods. Then the electrodes should be filed square so that they have sharp edges that promote proper spark formation. Finally, the gap is reset to specifications.

Spark plugs should ideally be tightened with a torque wrench. If that is not practical, tighten plugs with gaskets one-half turn after the gasket touches the head, and those with conical seats one-sixteenth turn after contact.

Aluminum heads are becoming more and more popular, and the thing to remember about them is always to use a suitable antiseize compound or graphite on the spark plug threads.

CARBS

Carburetors changed considerably in the antipollution era because they are so critical to the composition of what comes out of the tailpipe. Calibrations became leaner in every mode of operation, chokes were made to open faster, and adjustments were made differently. All this resulted in drivability problems and service misunderstandings.

One of the first things that you will notice on almost all emission-controlled carburetors is the presence of limiter caps on the idle mixture screws (Fig. 15–1). These reduce the amount of adjustment possible to keep the mixture from being set too rich or too lean. Of course, they are frequently removed by people who believe that they can achieve a smoother idle by enriching the mixture more than the caps allow.

MIX SETTINGS

Normally, acceptable idle smoothness can be attained within the limiter cap's range. If it must be removed, very specific procedures are published by the automakers for setting the mixture. Then new caps are supposed to be installed.

An infrared exhaust gas analyzer is sometimes required to fulfill the manufacturer's idle mixture setting recommendations. You are supposed to juggle the mixture screw and idle speed adjustments until certain levels of HC and CO are achieved at the tailpipe.

But another method of adjusting the blend has eclipsed the use of an infrared analyzer (although that expensive piece of equipment is still very useful in troubleshooting): propane enrichment. All that is required is a propane tank, an adjustable valve, a hose, and the specific service data for the vehicle at hand (Fig. 15–9).

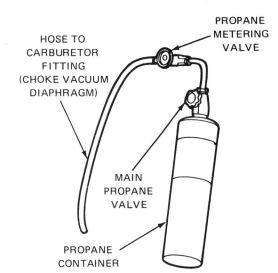

Figure 15–9 All that's needed to perform a propane enrichment idle mixture adjustment is a tachometer and this simple piece of equipment. (Chrysler)

PROPANE ENRICHMENT

We will use the instructions for a 1982 Chrysler 255-cubic inch six-cylinder model as an example of how a propane enrichment procedure is carried out. It should be used only if poor idle still exists after diagnosis has revealed no other faults, such

as incorrect ignition timing or
so on, or after a major carbı

1. Remove the idle mixtu
2. Connect a tachometer
   operating temperatur₍
3. Disconnect and plug t
   vacuum hoses.
4. Remove the air clea·
   nipple that goes to
   the nipple.
5. Remove the PCV
   then disconnect a
   canister.
6. With the engine
   rate is reached
7. With the propane still ı̇ʊⱳ.
   solenoid to obtain the propane-enricheu ı̇ₚ
8. Adjust the propane valve to see if you can raise the rpıı ı̇ꞇ.
   just the idle speed screw until you arrive once again at the propanc̫ꞇ̫
   speed.
9. Shut off the propane valve, then carefully adjust the idle mixture screw to achieve
   the smoothest idle at the specified hot idle speed.
10. Open the propane valve again and adjust it to obtain the highest idle speed.
    If it is now more than 25 rpm different from the enriched specification, repeat
    steps 6 through 9.
11. Shut off the propane, reinstall the vacuum hoses, replace the concealment plug,
    and perform the on- and off-solenoid speed and fast idle adjustments.

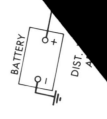

STOP DIESELING

The idle stop solenoid (Fig. 15–10) has been around since the late 1960s but is still
widely misunderstood, which is unfortunate because it is the most powerful deter-
rent to that annoying automotive phenomenon known as dieseling.

    This little device is attached to the carburetor throttle linkage. It consists of
an electromagnetic coil with a core or plunger that can move in and out. When energized, the plunger projects out from the coil and holds the throttle open to a certain
position. When deenergized, the plunger drops back inside the coil, allowing the throttle to close as much as the ordinary throttle stop screw permits.

    The solenoid is connected to the ignition switch, so that it is energized when
the key is on, keeping the engine at normal idle speed. When the key is switched off,
the plunger retracts, allowing the throttle to close more. Since this reduces the amount

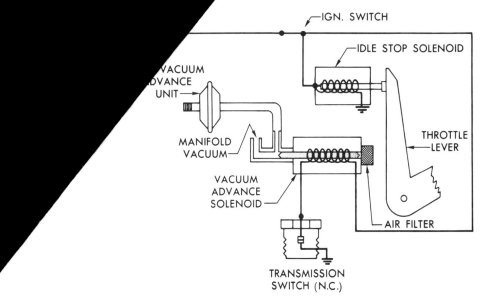

Figure 15-10   The idle-stop solenoid is energized when the ignition key is on so that it holds the throttle at the proper opening. The other system shown here is a transmission controlled spark advance. (Chevrolet)

of air that can enter the intake manifold, and the engine cannot run without air, it stops the engine dead even if self-ignition is occurring.

TWO IDLE SPEEDS

It is easy to set an idle stop solenoid properly (Fig. 15-11). First, find out both idle speed specifications—there will be one with the solenoid energized and another, lower one, with it deenergized. Set the former by either screwing the plunger in or out of the solenoid or by moving the solenoid itself. Then unplug the solenoid's electrical connector (which will cause the plunger to retract) and set the latter by means of the ordinary throttle stop screw.

Remember that the solenoid is not strong enough to overcome the throttle return spring, so when the key is switched off, then back on, or when the electrical connector is plugged back in, the throttle will have to be opened manually to allow the plunger to move out to its energized position.

KICKERS

Throttle *kickers* are devices that open the throttle a specified amount above normal idle under certain conditions, as when the air conditioner is switched on (Fig. 15-12). They may operate electrically or through a combination of electricity and vacuum.

Figure 15-11 On a car equipped with an idle-stop solenoid, always set curb idle speed with the solenoid, not the ordinary throttle stop screw. (General Motors)

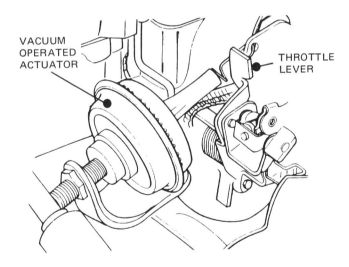

Figure 15-12 Throttle kickers can be used to keep idle speed constant in spite of accessory loads. (Ford)

Setting them properly requires only that you have the proper specifications for the vehicle.

A variation on this theme is the idle stabilizer found on Volkswagens with electronic ignition and other cars. The idea here is to maintain constant idle rpm, regardless of the accessories that are on, by advancing the timing. This, of course, is handled by a program in the ignition control module.

INJECTION CORRECTION

Then there is fuel injection, which is showing up on more and more cars. There are several types of electronic systems (EFI), but only one popular mechanical type—the Robert Bosch K-Jetronic (Fig. 15–13). The "K" stands for the word "constant" in German, so this setup is often called a Constant Injection System (CIS), and we will explain it briefly first.

A round sensor plate floating in a cone moves in relation to the amount of air rushing into the engine, and it is connected to a slit and piston valve that controls the fuel flow to the nozzles, one of which is aimed at each intake valve. They spray constantly, not intermittently as do those of a diesel or EFI system.

For years, CIS was entirely mechanical, but later models have added an oxygen sensor and an electronic control unit so that the mixture can be regulated accurately enough to accommodate a three-way catalyst (mixture variation is kept to an impressive 0.02 percent, where some ordinary carburetors can manage only 20 percent).

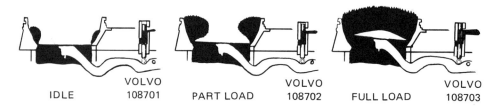

	VOLVO		VOLVO		VOLVO
IDLE	108701	PART LOAD	108702	FULL LOAD	108703

Figure 15-13 The Robert Bosch K-Jetronic CIS senses the volume of air entering the engine and mechanically adjusts the amount of fuel injected. (Volvo of America)

CIS SETTINGS

The two basic K-Jetronic adjustments that are involved in a normal tune-up are idle speed and mixture (Fig. 15–14). On a typical Volkswagen application, you will find the speed screw right next to the throttle plate linkage. The throttle is closed all the way at idle, and the screw varies the size of a bypass drilling—out equals faster.

The mixture is changed by turning a screw that bears on the lever that connects the sensor plate to the control plunger. There is a rubber plug between the fuel distributor and the air sensor cone. Remove it and insert a long 3-mm Allen wrench to rotate the screw. Clockwise rotation raises the percentage of CO. Late models with closed-loop/oxygen sensor systems have the access hole for the mixture adjustment sealed.

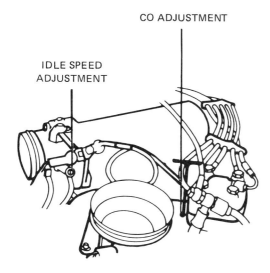

CO ADJUSTMENT

IDLE SPEED
ADJUSTMENT

Figure 15-14 The two basic K-Jetronic
CIS adjustments are shown here.
(Volkswagen of America)

ELECTRONICS ASCENDANT

Electronic fuel injection works on entirely different principles (Figs. 15-15 and 15-16).
Whether foreign or domestic, the idea is the same: Fuel is supplied to the injectors
(there may be one for each cylinder, one for each bank of vee-type engines, or only

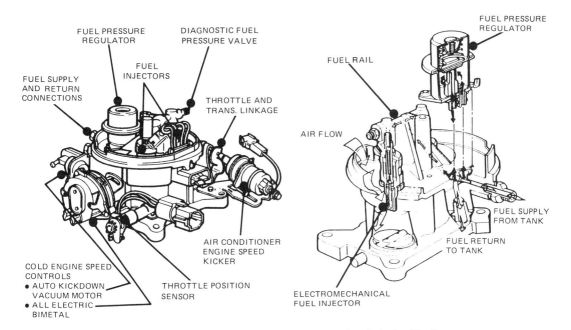

FUEL PRESSURE
REGULATOR

DIAGNOSTIC FUEL
PRESSURE VALVE

FUEL
INJECTORS

FUEL SUPPLY
AND RETURN
CONNECTIONS

THROTTLE AND
TRANS. LINKAGE

FUEL RAIL

FUEL PRESSURE
REGULATOR

AIR FLOW

AIR CONDITIONER
ENGINE SPEED
KICKER

FUEL SUPPLY
FROM TANK

FUEL RETURN
TO TANK

COLD ENGINE SPEED
CONTROLS
• AUTO KICKDOWN
  VACUUM MOTOR
• ALL ELECTRIC
  BIMETAL

THROTTLE POSITION
SENSOR

ELECTROMECHANICAL
FUEL INJECTOR

Figure 15-15 This is a domestic EFI throttle body. (Ford)

one for an in-line) at constant pressure by an electric pump. The injectors are actually solenoid valves and have only two positions: open or closed (Fig. 15–16). So the amount of fuel they feed into the engine is determined by the length of time they are open, called the *pulse width.*

An electronic control unit, actually a minicomputer, decides how long to make the electrical signal to the injectors and hence how much fuel is supplied on the basis of data it receives from various sensors. These usually report on manifold vacuum, coolant temperature, throttle position, engine rpm, and on closed-loop variations, the oxygen content of the exhaust. The Robert Bosch K-Jetronic system, which is used on many imports, also supplies the control unit with information on the volume of air entering the intake manifold by means of an airflow sensor that moves a variable resistor.

Some types have an injector aimed at each exhaust valve, which is called *port* or *multipoint injection,* whereas others place the injectors above the throttle plates, known as *single-point* or *throttle-body injection.*

Late-model domestic EFI systems incorporate a sophisticated self-diagnostic program into the computer. This will give anyone who is trying to fix a problem, information on the condition of the sensors, the control module, the wiring, and so on, in the form of codes.

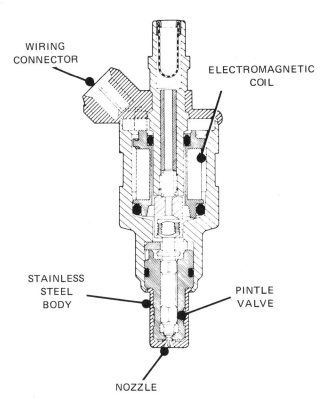

WIRING
CONNECTOR

ELECTROMAGNETIC
COIL

STAINLESS
STEEL
BODY

PINTLE
VALVE

NOZZLE

Figure 15–16 In all EFI systems, the injectors are actually solenoid-controlled fuel valves. (Ford)

Since there is a large number of different types of fuel injection systems, you will need specific service information for the vehicle at hand to troubleshoot the system, decipher the codes, or make adjustments.

SMOG FIGHTERS

Finally, there is the whole realm of emission control devices themselves. The PCV system should certainly be examined at every tune-up, and the valve replaced at least at the intervals specified.

There is always a good chance that some of those important vacuum hoses under the hood have gone hard, cracked, or have become disconnected or incorrectly routed, so check them during a tune-up.

An EGR valve that is stuck open will ruin the idle and reduce fuel mileage. Remove it for cleaning at the mileage suggested.

Beyond that, remember all the things that should be examined any time you open a hood—belts, hoses, and the condition of the coolant and the oil. When all of the above is added up, you can see that tune-ups will continue to be a large part of automotive maintenance, innovations notwithstanding.

16

Emission System Modifications

It has been estimated that a typical early 1970s car is about 15 percent less fuel efficient than it would be if it were not burdened with emission control devices. Along with this penalty, it is also prone to numerous drivability and performance complaints, and much more complicated to service than an equivalent car of the pre-emission-control era.

It should be obvious from the statements above that modification or elimination of antipollution equipment is attractive to many people. But there is a great deal more to the story. Legal, moral, and practical ramifications have to be considered, and the truth about the magnitude of any improvement that might or might not be realized must be investigated. That is the purpose of this chapter.

THE LAW

The legal situation is generally misunderstood, so we start with a discussion of that.

As far as the federal government and the Clean Air Act of 1970 are concerned, it is not illegal for a person to modify or even eliminate the emission control systems on his or her own car. One can do pretty much anything at all to one's own property and Uncle Sam will not be offended. In fact, there are no stipulations whatsoever regarding tinkering with your own automobile.

But just because Washington has not outlawed it, do not assume that your state does not say that it is illegal. Every state is required to meet federal ambient air quality standards, but how they do it is up to them. Some states have outlawed engine modifications that can affect emissions, while others have taken a different and, in our opinion, more reasonable route: They do not make any stipulations about what may be changed under the hood, but they check emissions at the tailpipe and demand that a repair be made if they are too high.

THE GARDEN STATE WAY

New Jersey's program is a good example of this. Every year, each car must visit a motor vehicle inspection station, which may be either a state-run emplacement or a licensed private garage. All safety-related items are checked—lights, wipers, brakes, horn, balljoints, tires, and so on—as they have been since New Jersey's PMVI program was started in the 1930s. But now there is an additional test that must be passed if the automobile is to be operated in that state: CO and HC emissions. As soon as the car enters the inspection lane, the probe of an infrared exhaust gas analyzer is inserted into its tailpipe. The standards vary according to the year of the vehicle (Fig. 16-1). They are, of course, less stringent for older models than for newer ones, the biggest step being between 1974 pre-catalytic-converter models and 1975 cars with the catalyst.

Since ordinary analyzers that are generally available and not apt to cause problems are used, only CO and HC are checked, not NO_x, which requires a different type of tester.

This program works to reduce air pollution without infringing too much on a car owner's freedom. Any engine modifications may be made as long as the actual tailpipe emissions are still within legal limits. On the other hand, certain changes, say, removing the catalytic converter, will almost surely cause the vehicle to fail inspection. So haphazard modifications that cause the car to become a gross polluter are effectively discouraged.

If you have any inclination to tamper with the emission controls on your car, be sure to check the laws of your home state to find out what is and what is not permitted, or you may be letting yourself in for citations, fines, or the inconvenience of failing inspection.

THE FEDS AND THE PROFESSIONAL MECHANIC

As we said above, the federal government has not outlawed modifications that you might perform on your own car. But it is very specific about what changes a professional mechanic working in a new-car dealership, service station, repair shop, or other service outlet may make on the emission control systems of a customer's car: none at all. Regardless of the state in which the mechanic works, it is illegal on the federal level for the person to tamper with the emission control devices with which the manufacturer equipped the car unless the changes are in the nature of a repair (Fig. 16-2). The penalties for violation are severe. Plugging even one vacuum hose for the purpose of deactivating an emission control system can result in a $10,000 fine.

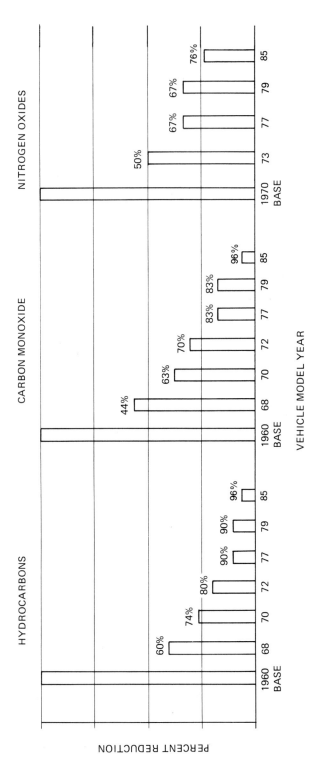

Figure 16-1 As these charts show, emissions controls have worked extremely well in reducing the levels of automotive air pollution. (Chrysler)

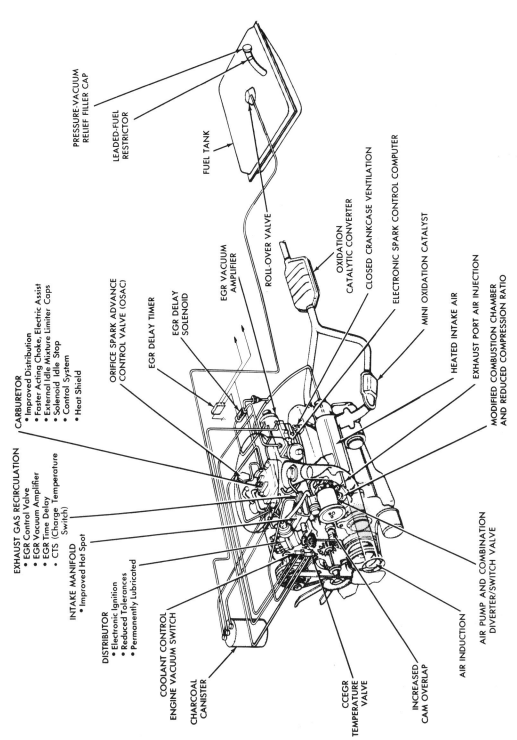

EXHAUST GAS RECIRCULATION
- EGR Control Valve
- EGR Vacuum Amplifier
- EGR Time Delay
- CTS (Charge Temperature Switch)

INTAKE MANIFOLD
- Improved Hot Spot

DISTRIBUTOR
- Electronic Ignition
- Reduced Tolerances
- Permanently Lubricated

COOLANT CONTROL
ENGINE VACUUM SWITCH

CHARCOAL
CANISTER

CCEGR
TEMPERATURE
VALVE

INCREASED
CAM OVERLAP

AIR INDUCTION

AIR PUMP AND COMBINATION
DIVERTER/SWITCH VALVE

CARBURETOR
- Improved Distribution
- Faster Acting Choke, Electric Assist
- External Idle Mixture Limiter Caps
- Solenoid Idle Stop
- Control System
- Heat Shield

ORIFICE SPARK ADVANCE
CONTROL VALVE (OSAC)

EGR DELAY TIMER

EGR DELAY
SOLENOID

EGR VACUUM
AMPLIFIER

ROLL-OVER VALVE

PRESSURE-VACUUM
RELIEF FILLER CAP

LEADED-FUEL
RESTRICTOR

FUEL TANK

OXIDATION
CATALYTIC CONVERTER

CLOSED CRANKCASE VENTILATION

ELECTRONIC SPARK CONTROL COMPUTER

MINI OXIDATION CATALYST

HEATED INTAKE AIR

EXHAUST PORT AIR INJECTION

MODIFIED COMBUSTION CHAMBER
AND REDUCED COMPRESSION RATIO

Figure 16–2 The emission control systems on today's cars are so complex and intertwined that any modifications you might make would almost surely result in drivability problems and a loss of efficiency. (Chrysler)

So anyone who works as a mechanic is risking a great deal by modifying engine control systems to improve the performance of a customer's car or to eliminate a drivability problem. The only wise way to service cars these days is to set everything to specifications and to repair or replace any antipollution devices that might be causing trouble. To do that, of course, it is necessary to learn about emission controls.

MORALITY/ETHICS

Certainly, there is always a good possibility of being able to get away with an illegal modification whether you are a mechanic or an ordinary motorist. But what about the moral burden you would be placing on yourself? We all have to live on the same planet and breathe the same air, so turning a car into a polluter can be construed as infringing on your neighbors' rights. The fact that we have a great deal of freedom in this country relative to most others requires that we take individual responsibility for our actions, that we try to be good citizens on our own without having a governmental watchdog to threaten us. By causing a car to emit more than it should, we are encouraging the creation of more and more laws that restrict our liberty.

Just think about it for a moment: If a large number of people eliminated the emission control systems from their cars, the states' ambient air quality standards would not be met. Therefore, stricter legislation would surely be enacted to put the states in compliance, and we would all suffer the effect of excessive government interference and bureaucratic confusion.

IS IT WORTH IT?

There is also the practical side, which can be stated quite succinctly: There is simply not that much to be gained in terms of performance and fuel economy by modifying or eliminating antipollution equipment, and it is very probable that you will cause more problems than you will eliminate.

According to one study, almost every amateur attempt to improve the efficiency of an engine by tampering with the emission control systems results in a loss of gasoline mileage. If you have a dynamometer-equipped laboratory and a staff of engineers and automotive technicians at your disposal, you can probably get some small incremental gains. But if you are working in your own garage, or under the tree in your backyard, the odds are definitely not in your favor. In most cases, you would be much better off to keep your car tuned to factory specifications and to make sure that all the antipollution systems are working as they should.

A LITTLE KNOWLEDGE. . . .

There are books on the market that explain how to modify or remove the antipollution devices on many cars, and they usually make claims of dramatic improvements in fuel mileage, performance, drivability, and even engine life. One, in particular, says on the cover that you can expect "15 to 40 percent better fuel economy." We find these statements to be dubious at best and downright irresponsible at worst. The book, which claims to be technically very accurate, also contains some misconceptions and leaves out many of the possible negative effects of the changes recommended. This is typical of the breed, which seems to be mostly a vehicle for selling rather unprofessional writing that skims the surface of a complicated subject.

The point is this: The chances are very good that you will realize little or no improvement by deactivating the emission control systems on your engine, and you are apt to cause new problems that will be difficult to diagnose and correct. Here, the adage "A little bit of knowledge is a dangerous thing" is certainly appropriate. The elimination of some antipollution devices will cause profound engine damage, while tampering with others will result in *reduced* drivability. So for both practical and moral reasons, the details of the changes mentioned hereafter are not in any way meant to encourage you to make them. On the contrary, it is our view, based on considerable experimentation, research, and general automotive repair experience, that the best way to get maximum efficiency and overall satisfaction from your car is to tune it up frequently using the specifications the manufacturer recommends and to learn about the emission controls present on it so as to keep them working properly.

The rest of this chapter describes modifications that can be made to each of the major emission control systems found on most cars; the potential gains, if any; and the drawbacks.

PCV

Positive crankcase ventilation was the first system to appear on cars sold in the United States and unless you want to ruin your engine, it is nothing to tamper with (Fig. 16–3). By using engine vacuum to draw crankcase gases into the intake stream, it vastly reduces the amount of sludge and varnish buildup on the internal components and eliminates the accumulation of water and acid in the oil. It maintains a flow of fresh air through the crankcase that picks up blowby and steam and meters them into the intake manifold to be burned in the cylinders.

Most pre-emission-control cars had a road-draft tube through which these gases exited into the atmosphere, but its action was nowhere near as positive as that of the PCV, so these older engines got much dirtier and more corroded inside than do post-control versions. In addition, a tiny bit of efficiency is gained by burning the HC in the blowby instead of just allowing it to escape. One of the authors did an experiment recently in which he removed the PCV setup on a 140-cubic-inch four-cylinder engine and vented the crankcase to the air cleaner housing to see if there would be enough negative pressure there to eliminate internal gases. After only 600 miles of highway driving, the vent hoses were clogged with sludge, the air filter element was soaked, and the inside of the engine had accumulated more deposits than it would have in many thousands of miles of driving with the PCV system intact.

So there is no justification whatsoever for deactivating the PCV on your car.

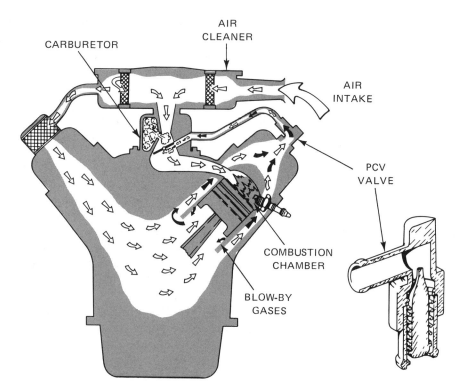

Figure 16-3 Disconnecting the PCV system is a sure way to clog up an engine. (Chrysler)

It does absolutely no harm to drivability, idle quality (provided that the valve is in good shape), and fuel mileage, and it does a lot of good. But it must be properly maintained in order to realize the benefits. Replace the valve at the recommended intervals, do the same with the little mesh intake filter that is usually found inside the air cleaner, and make sure that the hoses and passages are clear and undamaged.

VAPOR TRAP

The evaporative control system that collects the gasoline fumes that would normally escape into the atmosphere from the fuel tank and the carburetor bowl and then meters them into the intake stream appeared in 1971. This is another totally harmless and quite beneficial setup that should absolutely not be tampered with (Fig. 16-4). By collecting evaporating gasoline, it eliminates about 20 percent of the total emissions that come from a car, all the while saving fuel that you would lose without it. So leave it intact and make sure that the hoses, charcoal canister, and any filter that may be present are in good condition and connected properly.

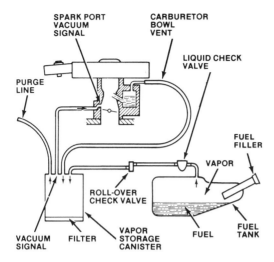

Figure 16-4 The evaporative emissions control system causes absolutely no loss of efficiency and even saves some gasoline. (American Motors Corp.)

INJECTING AIR

The air injection system [also called the air injector reactor (AIR) and the "smog pump" by mechanics], was, like the PCV, an early addition. By forcing fresh air into the exhaust ports or manifold, it gives the escaping HC and CO additional oxygen to combine with on its way through the exhaust system, effectively burning a considerable portion of it (Fig. 16-5). The only drawback to this system is the drag the pump produces. However, this is so miniscule that it usually cannot even be detected on a chassis dynamometer. The pump itself is a small vane-type unit, and if you disconnect the belt and spin it by hand, you will see how little power it requires. So there is not much to be gained by eliminating it.

On the other hand, in many applications the pump's pulley acts as an idler or adjustment point for a belt that also powers another accessory. In these cars, removal

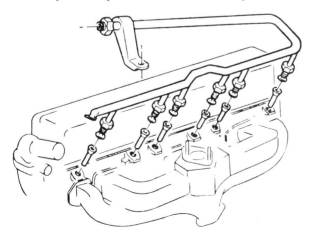

Figure 16-5 Air injection nozzles have virtually no effect on performance. (Chevrolet)

of the pump would require that a new idler pulley be fabricated and that is much more trouble than it is worth. Also, there is evidence that the increased temperatures present in the exhaust while the system is operating contribute to longer muffler and tailpipe life.

The small-diameter air injection tubes that project into the exhaust ports have been accused of restricting flow and sapping performance, so some people remove them and install pipe plugs in the holes. We have never seen a case where there was any increase in performance when this was done.

The aspirator valve system, like the Pulsair found on some General Motors cars, does basically the same thing as the air injection pump, although it has less capacity. It is also less complicated and has no pump to cause drag. This is yet another emission control device that does not cost the engine anything at all in terms of efficiency. There is no justification whatsoever for eliminating it. It helps lower the HC and CO levels in the air we all have to breathe without using one extra drop of gasoline in the process. Also, it takes up little room under the hood and requires no belt to drive it or any maintenance at all. Let it be.

EGR

The principle of the exhaust gas recirculation (EGR) system is that it admits a measured amount of exhaust into the intake stream to dilute the air/fuel mixture. This keeps the burn in the cylinders from reaching the peak temperatures that produce NO_x.

Ever since its introduction in 1972, many people have considered EGR to be one of the emission control devices that has the most effect on engine efficiency. They assumed that adding exhaust to the incoming charge will naturally reduce the power and efficiency of the engine, so they removed or deactivated the EGR valve and expected to reap dramatic gains in fuel mileage and performance (Fig. 16-6).

Theoretically, at least, some improvement is possible through the elimination of EGR, but, as we will explain, given real-world circumstances, such elimination usually causes problems and nets no benefits.

Figure 16-6 Disconnecting and plugging the EGR valve hose is probably the tamperer's favorite trick. Unfortunately, it can cause a loss of efficiency plus serious engine damage by promoting heavy detonation.

The most probable result of an inoperative EGR is detonation. The higher combustion chamber temperatures that occur without the dilution of the air/fuel mixture can cause the charge to explode violently (which is what makes the "pinging" sound) rather than burn smoothly as it should, imparting an even push to the top of the piston rather than a sharp blow.

Detonation is bad news. Not only do those constant shocking explosions damage the piston and rings, but the full work potential of the charge in the cylinder is not realized because the power is dissipated violently instead of being used to force the piston downward. Therefore, detonation can reduce fuel efficiency. With today's relatively low octane fuels, most cars operate right at the point of detonation anyway, and the removal of the EGR will probably push them over the edge.

One of the reasons this system is so frequently deactivated is that it is so simple to do. Just remove the vacuum line from the EGR valve nipple and plug the line. This will cause the valve to stay in the closed position and no recirculation will occur.

LATER

Ignition timing is often considered fertile ground for modification (Fig. 16–7). Especially on early emission-controlled engines, late spark was used to give the kind of burn that oxidized a great deal of HC and CO that would otherwise escape through the tailpipe, and it is true that a certain amount of efficiency was sacrificed to this.

If you have the facilities, patience, and accurate methods of testing the results, it is sometimes possible to improve fuel mileage and power somewhat by both advancing initial spark lead and recalibrating the advance curve on those older power plants. On the other hand, with today's gasoline, this type of modification may result

ORIFICE SPARK ADVANCE
CONTROL VALVE (OSAC)

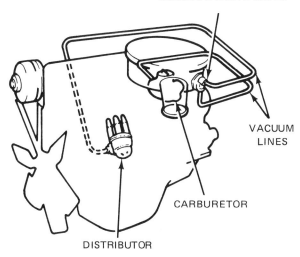

VACUUM
LINES

CARBURETOR

DISTRIBUTOR

Figure 16–7 Systems that delay spark advance, such as this OSAC (Orifice Spark Advance Control) set-up, are frequently bypassed. Usually, the only result is detonation during acceleration. (Chrysler)

in severe detonation. In fact, with some high-compression engines of the past, it is necessary to retard the spark to keep pinging within bounds with 89-octane fuel.

If you enjoy tinkering with advance springs and weights, you may want to experiment with the advance curve. But do not expect a dramatic improvement, and be prepared to put everything back the way it was if your efforts cause detonation. Also, if your state has an emission inspection program, there is a good chance that advancing the spark will cause the car to fail the test.

On later models, almost everything imaginable has been done to squeeze the very last bit of fuel economy out of the engine, so there is very little chance that you can gain anything at all from ignition timing modifications. In fact, you will pretty surely cause efficiency to drop. As you have probably noticed, most newer cars are prone to detonation even when tuned to specifications, so any increase in spark advance that you introduce will just make the problem worse.

CARBURETION

As far as improving fuel mileage is concerned, do not even bother thinking about carburetor modification. One of the first things the engineers did to combat tailpipe emissions was to lean out all the carburetor circuits as much as possible and still have the engine run, so you have no room to change anything.

The only worthwhile possibility here is the purchase of one of those newer super-efficient carburetors, such as Holley's Economaster line, with perhaps a high-velocity aftermarket intake manifold. These carefully engineered items can improve fuel economy, throttle response, and in some cases, overall performance while maintaining or even reducing the stock levels of emissions. They are, however, expensive and require considerable work for installation.

CAT ELIMINATION

In spite of the euphemism "test pipe," we all know what those lengths of exhaust tubing are for: the elimination of the catalytic converter. Apparently, this modification is done quite frequently. In one EPA case, a muffler shop in Kentucky was accused of removing 48 catalytic converters (the agency is seeking a $120,000 fine).

Just as with other emissions systems, removal of the catalyst is not, in our opinion, worth the trouble. A very slight improvement in mileage and performance might be gained by taking advantage of the two-point-higher octane rating that leaded regular gas usually has over unleaded. To do this, you remove the cat, install the "test pipe," punch out the restriction in the gasoline tank filler neck, and advance ignition timing two or three degrees.

The disadvantages of doing this include faster fouling of the spark plugs, contamination of the oil, and the inability to pass emissions inspection (if your state has such a program). Also, since it is illegal for a service station operator or employee to introduce leaded gasoline into a vehicle that is intended to run on unleaded only (the fine can be up to $10,000), you might have some trouble getting filled up.

Glossary

acid rain   One of the nasty effects of air pollution. NO_x from automobile exhaust and sulfur dioxide from the burning of high-sulfur-content coal and oil enter the atmosphere and undergo a series of complex chemical reactions. The end product is sulfuric acid. When it rains, the sulfuric acid is washed from the clouds and falls into lakes and streams, where it can build up to deadly concentrations. Many lakes in the eastern part of the United States have lost their fish due to acid rain. Acid rain is also harmful to buildings and painted finishes.

air   The air in our earth's atmosphere is composed of two main gases: nitrogen (76 to 78 percent) and oxygen (18 to 21 percent). The remaining few percent include carbon dioxide, argon, and other gases. When air is drawn into an engine, the oxygen combines with the fuel during combustion, producing carbon dioxide and water vapor. If there is too much fuel for the available oxygen, carbon monoxide and unburned hydrocarbons are produced in the exhaust. What is more, at temperatures above 2500 °F, oxygen can combine with nitrogen to form NO_x, another harmful pollutant.

air/fuel ratio   The relative proportions of air and fuel produced by the carburetor, fuel injection, or inside the intake manifold. The ideal or "stoichiometric" air/fuel ratio for gasoline is 14.7:1. A higher ratio would contain more air and less fuel, and would be considered a "lean" mixture. A lower ratio with more fuel and less air would be a "rich" mixture. The air/fuel ratio is determined by the orifice size of the main jets inside a carburetor, the dwell duration of the mixture control solenoid inside a feedback carburetor, or the orifice opening and fuel pulse duration of a fuel injector.

air injection By forcing air into the exhaust manifold, additional oxygen is mixed with the hot exhaust gases. When this mixture passes through a catalytic converter, the oxygen combines with the unburned hydrocarbons and carbon monoxide to form water vapor and carbon dioxide. The air is supplied by either an air pump or is sucked into the exhaust system through a PULSE air valve. This valve opens when negative pressure pulses occur in the exhaust, allowing fresh air to be drawn into the exhaust.

air pump This device pumps air into the exhaust manifold so that extra oxygen will be present to combine with the pollutants in the exhaust; also called a smog pump. The reaction takes place inside the catalytic converter.

ambient temperature A term used to describe the temperature of the air surrounding a vehicle. It means the same thing as room temperature or outside temperature.

back pressure The pressure that builds up inside the exhaust system. Some emissions control systems use a back-pressure sensor or diaphragm to monitor back pressure so that EGR flow can be increased when the engine is under maximum load (and producing maximum back pressure). Restrictions in the exhaust system can increase back pressure and adversely affect performance, emissions, and fuel economy. A bent or collapsed exhaust pipe, blocked muffler, or clogged catalytic converter can cause excess back pressure.

barometric pressure The pressure of the atmosphere as measured with a barometer. Barometric pressure changes with the weather and with altitude. Since it affects the density of the air entering the engine and ultimately the air/fuel ratio, some computerized emissions control systems use a barometric pressure sensor so that the spark advance and EGR flow can be regulated to control emissions more precisely.

black box A slang term used to describe an electronic control device such as the ignition module or computer. They are called black boxes because most people do not understand what goes on inside. Also, black boxes are usually sealed shut. If one malfunctions, it must be replaced as a unit rather than disassembled and repaired.

Btu British thermal units are a measure of heat energy. One Btu equals the amount of heat required to raise the temperature of one pound of water one degree Fahrenheit. The heat value of a fuel is usually given in Btu's per gallon or per pound.

carbon monoxide An odorless, colorless gas that can be fatal to human beings in extremely small doses; chemical formula: CO. It results from the incomplete combustion of any fuel containing carbon (gasoline, diesel fuel, alcohol, coal, wood, etc.). If additional oxygen is provided (as by an air pump) and allowed to combine with the CO (as happens inside a catalytic converter), carbon monoxide is transformed into harmless carbon dioxide (CO_2). Excessive CO emissions are caused by overly rich air/fuel ratios. The causes include restrictions in the air intake, a clogged air filter, a partially closed choke, a fuel-saturated carburetor float, too high a float level, leaky needle valve and seat, oversized carburetor jets, or internal fuel leaks.

catalyst   Some metals, including platinum, rhodium, and palladium, have the unique ability of speeding up chemical reactions. Since these metals are very expensive, thin coatings are applied to ceramic pellets or honeycomb inside the converter housing. When the hot exhaust gases pass through the converter, the catalyst allows oxygen from the air pump to "burn" the exhaust a second time. This removes the HC and CO pollutants from the exhaust. Catalysts are quite sensitive to lead, however. If leaded gasoline is used, the lead coats the catalyst and renders it useless.

catalytic converter   The bedpan- or muffler-shaped device that holds the catalyst. Located in the exhaust system just behind the head pipe from the exhaust manifold, the converter's job is to remove HC and CO from the exhaust. As the pollutants are burned inside the converter, temperatures reach 1200 to 1600 °F. Because of this, the converter shell is made of stainless steel and is surrounded by a heat shield.

centrifugal advance   A mechanical means of advancing spark timing with flyweights and springs. These weights are located inside the distributor (except on engines with computerized engine controls), under the rotor (GM window-type distributors), or under the point or pickup base plate. The size of the weights, the amount of spring tension, and the engine rpm rating determines the rate of advance.

charcoal canister   A small cylindrical or rectangular container that contains activated charcoal particles. The charcoal traps gasoline vapors from a vehicle's sealed fuel system. It is part of the evaporative emission control system and prevents gasoline vapors from entering the atmosphere. The canister is connected to the fuel tank and carburetor vents with hoses. When the engine is started, a small valve opens so that the vapors which have collected in the canister will be drawn into the engine and burned. Some canisters have a small filter on the bottom that must be changed periodically.

choke   A little flaplike valve in the top of a carburetor opening that restricts airflow. The choke's purpose is to increase the richness of the air/fuel ratio during starting. Choke problems are the primary cause of hard starting. If the choke is not closed when the engine is started, the air/fuel ratio will be too lean while the engine is cold. This results in a slow idle and stalling. If the choke does not open up at the correct rate as the engine warms up, the air/fuel ratio becomes too rich. This causes a rough idle, poor fuel economy, and excessive HC and CO emissions. The choke is controlled by a coiled bimetallic spring that reacts to temperature changes. An electric heating element inside the choke housing, or warm exhaust gases or engine coolant routed nearby, are used to speed up the rate at which the choke opens.

Clean Air Act   Passed in 1970 by the U.S. Congress, this was the legislation that created today's auto emissions laws and established the Environmental Protection Agency as the watchdog over our nation's air quality.

compression ratio   The relationship between the piston cylinder volume from bottom dead center to top dead center. Higher compression ratios improve combustion efficiency but also require higher-octane fuels. Pre-emission-control engines often had compression ratios as high as 11:1, whereas most of today's engines are between 8.5:1 and 9:1. Diesel engines have very high compression ratios—from 18:1 to 22:1.

computerized engine controls   With General Motors' C-4 and Computer Command Control, and with Ford's Electronic Engine Control, a small microprocessor is used to control both spark timing and the carburetor's air/fuel ratio. The computer allows much more precise control of these functions than do mechanical or vacuum devices. Sensors tell the computer about engine temperature, rpm, inlet air temperature, throttle position, and so on, so that it can calculate how much spark advance is needed for optimum performance and minimum emissions. An oxygen sensor in the exhaust manifold tells the computer about the oxygen content in the exhaust gases so that the computer can control the air/fuel mixture solenoid inside the carburetor. Computerized engine controls are self-diagnostic to a certain extent, and will store fault codes in their memory to help the mechanic diagnose any problems that may exist.

crankcase emissions   Since piston rings do not do a perfect job of sealing cylinders, a small amount of combustion gases will blow by the rings and enter the engine's crankcase with every revolution of the motor. These gases include unburned gasoline, partially burned gasoline and oil, water vapor, and other pollutants. In pre-emission-control days, these pollutants were allowed to vent to the atmosphere through breather tubes. In the mid-1960s, however, positive crankcase ventilation (PCV) was introduced. The crankcase was sealed so that these gases would be drawn through a small valve and into the intake manifold to be reburned in the engine. This eliminated a major source of pollution and improved crankcase ventilation, since engine vacuum continually sucked the pollutants out before they could accumulate and cause lubrication problems.

DEFI   An abbreviation for "digital electronic fuel injection." This system is used by Cadillac. It uses a microprocessor to control both spark timing and throttle body fuel injection. The fuel injectors are pulsed electronically to meter fuel flow.

detonation   Caused by fuel burning erratically or explosively inside the combustion chamber; also called spark knock or pinging. Detonation can be caused by excess spark advance, low-octane fuel, lean air/fuel mixtures, and/or overheating. Carbon buildup inside the combustion chambers and on the piston face can also increase compression sufficiently to cause detonation. Mild detonation is not harmful, but heavy detonation can break piston rings, burn holes through the tops of pistons, and damage rod bearings. Detonation also causes elevated HC emissions in the exhaust.

dieseling   A term used to describe engine run-on or preignition. The engine continues to cough and run after the ignition has been switched off because hot spots inside the combustion chambers are spontaneously igniting the incoming fuel. Sharp edges and carbon deposits inside the combustion chambers aggravate an engine's tendency to diesel. The higher operating temperatures of today's emission-controlled engines and low-octane fuels add to the problem. An idle stop solenoid at the carburetor prevents dieseling by completely blocking the flow of air and fuel into the engine. If an engine diesels, check for a faulty or misadjusted idle stop solenoid.

diverter valve   Part of the air injection system, the diverter valve diverts air from the air pump either to the exhaust system or to the atmosphere, depending on the need. The diverter valve provides pressure relief whenever the air pump volume is excessive. On cars with three-way catalytic converters, the valve can also supply additional air to the second catalyst.

dynamometer   A machine used to measure engine horsepower output. An engine dynamometer measures power output at the crankshaft and a chassis dynamometer measures power output at the driving wheels.

EGR valve   The exhaust gas recirculation valve meters a small amount of exhaust gas into the intake manifold to dilute the air/fuel mixture. This keeps combustion temperatures below 2500 °F to reduce the formation of NO_x. The amount of exhaust gas recirculated into the engine is only a few percent. The EGR valve is mounted on the intake manifold. A vacuum-operated diaphragm lifts a small plunger-type valve that uncovers an opening to the intake manifold. Exhaust gases then flow through plumbing from the exhaust manifold past the EGR valve and into the engine.

electronic spark control   The process whereby spark advance is controlled electronically rather than by mechanical means. A microprocessor calculates the necessary amount of spark advance based on engine operating conditions and what has been programmed into its memory. The traditional vacuum and centrifugal advance mechanisms are not used in the distributor.

EPA   The Environmental Protection Agency is the federal government's agency that determines air quality standards, enforces those standards, and certifies new automobiles as meeting those standards. A fine of up to $10,000 can be imposed on any auto repair establishment that intentionally disconnects or renders inoperative any emissions control device. It is not against the law, however, for a person to tamper with the pollution controls on his or her own automobile.

evaporative emissions   This refers to the gasoline fumes that evaporate from the fuel tank and carburetor bowl. These fumes contribute to smog problem when allowed to escape into the atmosphere. They are eliminated by sealing the fuel system and using a charcoal canister to trap vapors from the fuel tank and carburetor. California also requires that filling stations include evaporative emission control devices on gasoline pumps to catch gasoline fumes when the fuel tanks are being filled.

exhaust emissions   The principal pollutants the EPA is concerned with in automobile exhaust emissions are lead, unburned hydrocarbons (HC), carbon monoxide (CO), and oxides of nitrogen (NO_x). Permissible levels for these pollutants in terms of grams per mile or percent of volume are regulated by law.

exhaust gas recirculation   The process of recycling burned exhaust gases into the intake manifold to dilute the air/fuel mixture and reduce combustion temperatures. Since the exhaust gases have already been burned, they are inert. This keeps combustion temperatures below 2500 °F to minimize the formation of oxides of nitrogen (NO_x), a pollutant that is the primary contributor to smog.

feedback carburetion   Used in conjunction with computerized engine control, an air/fuel mixture solenoid is linked to the computer to constantly modify the air/fuel ratio according to need. An oxygen sensor in the exhaust manifold tells the computer how much oxygen is left in the exhaust after combustion. The computer then changes the duration of its electronic signal to the mixture control solenoid to increase or decrease the richness of the air/fuel ratio. Feedback carburetors make a buzzing sound and can be diagnosed with a dwell meter.

fuel injection   A system that uses no carburetor but sprays fuel under pressure into the intake manifold or directly into the cylinder intake ports. The advantage of fuel injection over carburetion is that it allows more precise control of the air/fuel mixture for improved performance, fuel economy, and reduced exhaust emissions.

gulp valve   Senses the increase in manifold vacuum caused by the rapid closing of the carburetor throttle plates during deceleration; also called a decel valve. Under such circumstances, the valve opens and allows filtered air to enter the intake manifold. This leans out the air/fuel mixture to prevent engine backfire in the exhaust and to reduce HC and CO emissions.

HC-CO meter   Uses an infrared sensing device to measure the amount of HC and CO in vehicle exhaust; also called an exhaust gas analyzer. A probe is inserted into the tailpipe of the vehicle being checked so that samples of the exhaust gases can be drawn into the machine. HC is read in parts per million and CO is read in percent.

heated-air system   The system that routes heated air from around the exhaust manifold into the air cleaner snorkel during engine warm-up. The system uses a thermostatically controlled flap in the air cleaner snorkel to block off cold-air intake. Warm air is needed during engine warm-up and cold weather to vaporize the fuel properly. A malfunction of the system that prevents the door from closing will lead to hesitation and stumbling while the engine is cold. An air temperature control flap stuck shut will overheat the air/fuel mixture, possibly causing detonation and elevated CO levels (due to a rich air/fuel ratio, as warm air is less dense than cold air).

heat riser valve   A control valve that fits between the exhaust manifold and exhaust pipe on one side of a V8 or V6 engine. The valve restricts the flow of exhaust while the engine is cold to speed engine warm-up. The gases circulate back through a passage in the intake manifold to the opposite side of the engine. This warms up the intake manifold and carburetor to help vaporize droplets of gasoline in the air/fuel mixture. A heat riser valve stuck open will slow engine warm-up and may cause hesitation and stalling while the engine is cold. A valve stuck in the closed position will greatly restrict the exhaust system and cause a noticeable lack of power and drop in fuel economy.

hot idle compensator   A temperature-sensitive carburetor valve that opens when the inlet air temperature exceeds a certain level. This allows additional air to enter the intake manifold to prevent overly rich air/fuel ratios.

humidity   The amount of water vapor in the air. The amount of water air can hold before it becomes saturated depends on temperature. Warm air can hold more moisture than cold air. At 100 percent humidity, the air is completely saturated with moisture. Humidity affects engine performance because it tends to boost the effective octane rating of the air/fuel mixture. Engines can therefore tolerate more spark advance during humid weather than during dry weather.

hydrocarbons   Any compound containing hydrogen and carbon atoms. There are many different types of hydrocarbons, including crude oil, gasoline, diesel fuel, kerosene, and even coal. Unburned hydrocarbons include gasoline vapors, partially burned gasoline and oil vapors, and soot. HC is one of the major air pollutants.

idle adjustment   Can refer to either the idle speed adjustment or the idle mixture adjustment. The idle speed is set by turning a screw that opens the carburetor throttle plates. The screw is turned until the desired idle rpm level is achieved. The idle mixture adjustment is set by turning a screw that opens and closes a port in the carburetor. Turning the screw out richens the idle mixture, while turning it in leans the mixture. Using the idle drop technique, the mixture is set by adjusting the mixture for smoothest idle at the slowest rpm rating. On late-model engines, the propane enrichment technique must be used to adjust idle mixture.

idle limiter   A plastic cap with tabs that limits idle mixture adjustments to a pre-determined range. This is to prevent overly rich or lean mixture adjustments that could cause increased exhaust emissions.

idle mixture   The air/fuel ratio of the idle circuit of the carburetor. The mixture is determined by turning an adjustment screw. On some late-model carburetors, the idle mixture screw is set at the factory and then sealed to prevent tampering.

idle speed   The speed at which the engine idles. For most engines, the idle speed is somewhere between 600 and 850 rpm. A set of cam lobes on the choke linkage provides a fast idle speed during engine warm-up. The idle speed is adjusted by turning a screw that opens the throttle plates. On some feedback carburetors and fuel injection systems, the computer controls the idle speed.

I/M   Abbreviation for "inspection and maintenance." This is the periodic inspection of a vehicle's emission control system, ignition, and carburetion with the intention of maintaining it in good working condition. Many states have I/M programs whereby motorists must have their vehicles inspected yearly for emissions and safety certification.

magnetic timing   Checking or adjusting an engine's ignition timing using a mag timing meter rather than a timing light. A magnetic probe is inserted into a receptacle near the crankshaft harmonic balancer or flywheel. The probe picks up a small notch in the balancer or flywheel with every revolution of the engine. An inductive pickup that clamps onto the number one plug wire tells the meter when the plug fires. The meter then displays the degrees of timing advance.

manifold vacuum   The vacuum created inside the intake manifold by the intake strokes of the pistons. Manifold vacuum is measured by a vacuum gauge that reads inches or millimeters of mercury. Manifold vacuum can be used to diagnose a variety of ills inside an engine.

monolith converter   "Monolith" means made of a single unit. With respect to the converter, it refers to the use of a single honeycomb structure for the catalyst as in most Ford catalytic converters. General Motors, on the other hand, uses coated pellets in their converters.

NIASE   An abbreviation for the "National Institute for Automotive Service Excellence." This organization certifies professional mechanics on a voluntary basis.

NO_x   An abbreviation for "nitrogen oxides" or "oxides of nitrogen." This is a pollutant that is formed inside the engine when combustion temperatures exceed 2500°F. At this temperature, the nitrogen in the air combines with oxygen to form a variety of toxic gases. NO_x formation is minimized by the use of exhaust gas recirculation.

octane   This refers to gasoline's ability to resist detonation. The higher the octane number, the greater the fuel's resistance to detonation. Most of today's gasolines have a pump octane rating of 87 to 92.

OSAC valve   An abbreviation for "orifice spark advance control valve," a device used on some Chrysler engines to limit NO_x formation. The valve delays vacuum to the distributor vacuum advance between idle and part throttle operation.

oxidation   The combining or burning of any material with oxygen. Rust is the slow oxidation of iron. The explosion of the air/fuel mixture is the rapid oxidation of gasoline. The combining of unburned HC with additional oxygen in the catalytic converter is oxidation of the unburned hydrocarbons.

oxygen sensor   A sensor in computerized engine control systems that is screwed into the exhaust manifold so that it can read the amount of oxygen left in the combustion gases. The sensor changes its output voltage in proportion to the amount of oxygen in the exhaust. The signal is fed to the computer so that it can control the air/fuel mixture at the carburetor for optimum performance and minimum emissions.

particulates Small particles of lead in the exhaust that result from burning leaded gasoline. Particulate pollution is regulated by so many grams per mile. Also refers to the soot particles from gasoline and diesel fuels. Particulate emissions of diesel engines are notoriously high and pose a significant health hazard in congested areas. The soot particles are so small that they enter the lungs and accumulate over time. The eventual result is an increased tendency to develop lung cancer.

PCV valve Abbreviation for "positive crankcase ventilation valve," a device used to limit crankcase emissions by drawing crankcase vapors back into the intake manifold.

PMVI Abbreviation for "periodic motor vehicle inspection" (same as I/M). Many states have PMVI programs whereby motorists must have their vehicles inspected yearly for emissions and safety violations. The object of PMVI is to make sure that emission control systems are working to ensure air quality.

pollutants Any substance that is harmful to the environment. With respect to automobiles, these include HC, CO, NO_x, lead, evaporative emissions, and crankcase emissions.

ported vacuum By positioning small openings or ports in the carburetor throat above the throttle plates, vacuum will not be created within the ports until the throttle plates open past the ports. Ported vacuum is used to control EGR valve operation and other vacuum-dependent devices.

positive crankcase ventilation PCV is used to reduce crankcase emissions. Combustion blowby gases are drawn through the PCV valve into the intake manifold for reburning in the engine.

preignition The ignition of the air/fuel mixture in the combustion chamber by means other than the spark; the same as dieseling. It is usually caused by hot spots in the combustion chamber (sharp edges or carbon accumulation). It can cause an engine to run-on after the ignition is switched off, and can cause engine damage if severe enough.

propane enrichment Using propane to make idle mixture adjustments on late-model carburetors. Propane is fed into the intake manifold at a controlled rate through a tube connected to a vacuum fitting. This helps to compensate for the lean air/fuel calibration during idle mixture adjustment.

retard The opposite of advance. With respect to ignition timing, retarding the timing is taking away degrees of spark advance. Spark timing must be retarded during certain conditions to prevent detonation, excess HC, and NO_x. Some distributors use a dual-diaphragm vacuum advance. One side advances while the other retards.

rich mixture An air/fuel mixture that has more than the usual amount of fuel in it. The ideal air/fuel ratio for gasoline is 14.7:1 by weight. Mixtures with less air and more fuel are considered rich (11:1, 9:1, 8:1, etc.). A rich mixture will increase engine power up to a point, but it also increases CO emissions.

run-on   The engine continues to run after the ignition has been switched off due to the spontaneous combustion of fuel; the same as dieseling. Run-on is usually due to hot spots inside the combustion chamber, low-octane fuel, or a misadjusted or inoperative idle stop solenoid.

sensor   An electromechanical device that is used with computerized engine control to monitor various engine functions or operating conditions. The sensor tells the computer how many the rpm rate at which the engine is turning, the temperature of the surrounding air, the content of oxygen in the exhaust, whether or not the throttles are closed, and so on, so that the computer can calculate the best air/fuel ratio and amount of spark advance.

smog   A term used to describe photochemical air pollution that results from exhaust pollutants, particularly NO_x and HC interacting with sunlight. Sunlight causes the chemicals to react to form ozone and a brownish nitrous oxide haze in the atmosphere. Smog can be blamed on automobile exhaust; stationary sources of pollution, such as steel mills, coal-burning utility plants, factories, and so on; and on natural sources that emit HC, NO_x, and sulfur dioxide.

spark advance curve   The rate at which the ignition timing advances. If plotted on a graph, the line would resemble a curve. It would rise from some initial amount of advance and finally level off at the maximum advance. On conventional ignition systems, the curve depends on the vacuum and centrifugal advance mechanisms. On computerized systems, the curve is programmed into the memory of the computer.

spark decel valve   A vacuum valve located in the line between the distributor and carburetor. The valve advances spark advance during deceleration to reduce emissions.

spark delay valve   Another type of vacuum valve that is sometimes used in the line between the distributor and carburetor. The delay valve delays vacuum advance under certain conditions to limit NO_x formation. The valve acts like a restriction in the vacuum line so that vacuum builds up and changes very slowly.

spark knock   Detonation or pinging caused by overadvanced ignition timing. On some late-model computerized engine control systems, a knock sensor is used to detect spark knock. When the sensor picks up the detonation pulses, the computer automatically retards the ignition timing until the knock disappears.

tetraethyl lead   A lead compound which, when added to gasoline, increases the octane rating of the fuel. Unfortunately, lead is harmful to catalytic converters and cannot be used in cars designed to operate on unleaded fuel. Tetraethyl lead is used in regular gasoline, although in smaller amounts compared to several years ago.

Thermactor   Ford's name for its air injection system.

three-way converter   Sometimes called a dual catalytic converter, it uses two different catalysts to control HC, CO, and NO_x. Two-way converters used up to 1980 controlled only HC and CO.

throttle body fuel injection   A fuel injection system wherein the fuel injectors are located in a common throttle body. The throttle body looks like a carburetor from the outside, and sits in the usual position on the intake manifold. The advantage of throttle body fuel injection over other forms of fuel injection is that it is cheaper to build and easier to control.

transmission vacuum switch   Abbreviated TVS, it is a simple vacuum switch that prevents distributor vacuum advance when the transmission is in any gear but high. General Motors calls its version transmission-controlled spark, and Ford calls its version transmission-regulated spark.

vacuum advance   Advancing or retarding the ignition timing via a vacuum diaphragm linked to the distributor base plate. Vacuum advance allows the ignition timing to vary with respect to throttle opening and engine load.

vacuum amplifier   A device that boosts the strength of a vacuum signal so that there is sufficient vacuum in a line to operate some other device.

vacuum delay valve   A device that restricts the flow of vacuum through a vacuum line so that there is a gradual buildup of vacuum. A delay valve will also smooth out fluctuations in a vacuum signal.

vacuum motor   A small vacuum cylinder with a diaphragm that is linked to a control mechanism of some sort. A small vacuum motor on the air cleaner snorkel opens and closes the warm-air door inside the snorkel. Vacuum motors are also used to open heat riser valves, cold-air ducts, and headlight housings on some cars.

vapor-liquid separator   A part of the evaporative emissions control system that prevents liquid gasoline from flowing through the vent line to the charcoal canister.

vapor lock   A condition where excessive engine heat has caused the fuel in the fuel line or fuel pump to boil. The bubbles block the flow of fuel to the carburetor and prevent the engine from starting.

vapor recovery system   Part of the evaporative emissions control system that prevents gasoline vapors from escaping to the atmosphere. The vapors are trapped in the charcoal canister, and are then drawn into the engine and burned when the engine is started. Vapor recovery can also refer to preventing gasoline vapors from entering the atmosphere when a car is being refueled.

venturi   The narrow part of the carburetor throat. When air passes this point, the restriction causes it to speed up. This creates a partial vacuum that can be used to draw gasoline into the airstream.

APPENDIX A

Automotive Acronyms

General

A/C	air conditioning
AC	alternating current
ATC	after top center
ATDC	after top dead center
ATF	automatic transmission fluid
BTC	before top center
BTDC	before top dead center
BTU	British thermal units
CID	cubic inch displacement
DC	direct current
DOHC	dual overhead cams
F.I.	fuel injection
HP	horsepower
LED	light-emitting diode
MPG	miles per gallon
OE	original equipment
OEM	original equipment manufacture
OHC	overhead cam
PSI	pounds per square inch
RPM	revolutions per minute
TDC	top dead center
VAC	volts of alternating current
VDC	volts of direct current

VIN   vehicle identification number

Pollutants

CO   carbon monoxide

HC   hydrocarbons

NO_x oxides of nitrogen

Basic pollution control systems

AIR   air injection reaction

ECS   evaporation control system (Chrysler)

EEC   evaporative emissions control (Ford)

EFE   early fuel evaporation control system (GM)

EGR   exhaust gas recirculation

PCV   positive crankcase ventilation

Manufacturer's emissions control systems

C–3   Computer Command Control (GM)

C–4   Computer Controlled Catalytic Converter (GM)

CAS   Clean Air System (Chrysler)

CCC   Computer Command Control (GM)

CCS   Controlled Combustion System (GM)

CEC   Combustion Emission Control (GM)

EEC   Electronic Engine Control (Ford)

ELB   Electronic Lean Burn (Chrysler)

IMCO   Improved Combustion System (Ford)

MISAR   Microprocessed Sensing and Automatic Regulation (GM)

SCS   Series Controlled System (GM)

Ignition, electronics, and computers

BID   breakerless inductive discharge (AMC)

CCC   combustion control computer (Chrysler)

CDI   capacitor discharge ignition

ECA   electronic control assembly (Ford)

ECCS   electronic concentrated engine control system (Nissan)

ECM   electronic control module (GM)

ECU   electronic control unit (Ford)

HEI   high-energy ignition (GM)

MCU   microcomputer unit (AMC)

MCU   microprocessor control unit (Ford)

PROM   program unit for GM ECM

SCC   spark control computer (Chrysler)

SSI   solid-state ignition (Ford and AMC)

Spark control systems and devices

CCEVS   coolant-controlled engine vacuum switch (Chrysler)

CSC   coolant spark control (Ford)

CSSA   cold-start spark advance (Ford)

CTAV   cold-temperature-actuated vacuum (Ford)

CTO   coolant temperature override switch (AMC)

DRCV	distributor retard control valve
DSSA	dual signal spark advance (Ford)
DVDSV	differential vacuum delay and separator valve
DVDV	distributor vacuum delay valve
ESA	electronic spark advance (Chrysler)
ESC	electronic spark control (Chrysler)
ESS	electronic spark selection (Cadillac)
EST	electronic spark timing (GM)
OSAC	orifice spark advance control (Chrysler)
PVA	ported vacuum advance
PVS	ported vacuum switch
SAVM	spark advance vacuum modulator
SRDV	spark retard delay valve
TAV	temperature-actuated vacuum
TIC	thermal ignition control (Chrysler)
TCS	transmission-controlled spark (GM and AMC)
TRS	transmission-regulated spark (Ford)

Sensors and switches

BARO	barometric pressure sensor (GM)
B-MAP	barometric/manifold absolute pressure (Ford)
BP	back-pressure sensor (Ford)
BPT	back-pressure transducer
CESS	cold engine sensor switch
CP	crankshaft position sensor (Ford)
CTS	charge temperature switch (Chrysler)
CTS	coolant temperature sensor (GM)
CTVS	choke thermal vacuum switch
ECT	engine coolant temperature switch (Ford)
EGO	exhaust gas oxygen sensor (Ford)
EPOS	EGR valve position sensor (Ford)
EVP	EGR valve position sensor (Ford)
IAT	inlet air temperature sensor (Ford)
MAP	manifold absolute pressure sensor (Ford)
MAT	manifold air temperature sensor (GM)
MCT	manifold charging temperature (Ford)
OS	oxygen sensor (GM)
TP	throttle position sensor (Ford)
TPP	throttle position potentiometer
TPS	throttle position sensor (GM)
TPT	throttle position transducer (Chrysler)
TVS	thermal vacuum switch (GM)
VSS	vehicle speed sensor
WOT	wide-open-throttle switch (GM)

Fuel system

AIS	automatic idle speed motor (Chrysler)
ASD	automatic shutdown module (Chrysler)
CANP	canister purge solenoid valve (Ford)
CCIE	coolant-controlled idle enrichment (Chrysler)
CEAB	cold-engine air bleed
CER	cold-enrichment rod (Ford)
CVR	control vacuum regulator (Ford)
DEFI	digital electronic fuel injection (Cadillac)
EFC	electronic feedback carburetor (Chrysler)
EFC	electronic fuel control
EFI	electronic fuel injection
FBC	feedback carburetor (Ford)
FBCA	feedback carburetor actuator (Ford)
FCA	fuel control assembly (Chrysler)
FCS	fuel control solenoid (Ford)
FDC	fuel deceleration valve (Ford)
ISC	idle speed control (GM)
ITS	idle tracking switch (Ford)
M/C	mixture control solenoid (GM)
MCS	mixture control solenoid (GM)
PECV	power enrichment control valve
TBI	throttle body injection
TIV	Thermactor idle vacuum valve (Ford)
TKS	throttle kicker solenoid (Ford)
TSP	throttle solenoid positioner (Ford)
TV	throttle valve

EGR and exhaust related

BPV	back-pressure valve
CC	catalytic converter
CCEGR	coolant-controlled EGR valve (Chrysler)
COC	conventional oxidation catalyst
EGR	exhaust gas recirculation
ORC	oxidation-reduction catalyst (GM)
TVV	thermal vacuum valve
TWC	three-way catalyst (Ford)

AIR-system related

AIR	air injection reactor
TAB	Thermactor air bypass solenoid (Ford)
TAD	Thermactor air diverter solenoid (Ford)

Miscellaneous

CAFE	corporate average fuel economy
CARB	California Air Resources Board

CATF   California Automotive Task Force
EPA   Environmental Protection Agency
PMVI   Preventative Maintenance Vehicle Inspection
NIASE   National Institute for Automotive Service Excellence (ASE)

Index

A

Air injection:
  air pump, 50–52
  aspirator valve, 56–59, 194
  description, 48–59, 157, 193–194
  diverter valve, 51–53, 69, 157
  gulp valve, 53–54
  thermal reactor, 55
  Pulsair, 58, 158
  vacuum differential valve, 53

B

Back pressure, 74–77

C

CARB, 7
Carbon monoxide:
  causes of, 121–128, 134, 136–139, 153–154, 159
  control of, 14, 61, 65, 69–71, 99–100
  description, 15–17
  formation of, 13
  testing for, 129–142, 149, 153–154, 187
Carburetion:
  air/fuel ratio, 13, 41, 67, 88–89, 134–135, 143, 150–151, 154

basics of, 88–99, 109, 143, 179–181
choke, 94–95, 153, 165
feedback, 67–68, 97, 155–156
fuel injection, 156, 182–185
idle mixture adjustment, 91–92, 100, 145, 153, 171, 178
modification of, 196
propane enrichment, 100, 145, 178–179
stoichiometric, 13, 67, 88–89
troubleshooting, 98–100, 143, 153–154, 161, 164–168
variable venturi, 97–98
Catalyst, 61–63
Catalytic converter, 55–58, 60–69, 138, 196
Centrifugal advance, 105–107
Charcoal canister, 36–39, 192
Clean Air Act, 4, 60, 186
Combustion, 13–14, 101–103, 121–128
Computerized engine controls, 56–57, 79–81, 97, 116–120, 170
Crankcase emissions, 12, 20–22

D

Detonation, 71, 104, 127, 138, 169, 196
Dieseling, 168, 179–180
Diverter valve, 51–53, 69, 157

E

EGR valve, 72–81, 84–86, 166, 194–195
Emission controls:
   air injection, 48–59, 157, 193–194
   air pump, 50–52
   catalytic converter, 55–58, 60–69, 138, 196
   charcoal canister, 36–39, 192
   EGR valve, 72–81, 84–86, 166, 194–195
   exhaust gas recirculation, 14, 70–87, 138,
      194–195
   heated air intake, 40–47, 153, 167, 169
   PCV valve, 24–25, 157, 168
   positive crankcase ventilation, 12, 20–28,
      124–125, 138, 144, 157, 191–192
Emission standards, 1–8, 188
Emissions testing, 129–162, 187
Evaporative emissions, 11–12, 29–39, 192
Exhaust gas recirculation, 14, 70–87, 138,
   194–195

F

Four gas analysis, 130–135, 139–140, 161

G

Gas cap, 32–33
Gulp valve, 53–54

H

HC-CO meter, 99, 129–139, 187
Heated air system, 40–47, 153, 167, 169
Hot idle compensator, 94, 96
Hydrocarbons:
   causes of, 99, 121–128, 150–153, 159
   control of, 61, 69–71
   description, 15–16
   formation of, 13
   testing for, 99, 129–142

I

Idle limiter, 172
Ignition system:
   centrifugal advance, 105–107
   computer advance, 116–120, 174
   description, 101–120
   OSAC valve, 114
   spark delay valve, 113–116

   timing, 103–108, 141, 165, 173, 195–196
   transmission vacuum switch, 78–79
   vacuum advance, 107–116

L

Liquid-vapor separator, 34–35

M

Magnetic timing, 174

N

NOX:
   control of, 65–66, 70–87, 126, 194
   description, 14–15, 17
   formation of, 14, 69
   testing for, 140, 160

O

Octane, 127, 196
OSAC valve, 114
Oxidation, 63
Oxygen sensor, 68, 97

P

Particulates, 16
PCV valve, 24–25, 157, 168
Pollutants, 11–19
Ported vacuum, 73, 109
Ported vacuum switch, 114
Positive crankcase ventilation, 12, 20–28,
   124–125, 138, 144, 157, 191–192

R

Reduction, 65
Retard, 105

S

Sensor, 68, 97, 116–119
Smog, 11, 16
Spark advance, 105–116
Spark delay valve, 113–116

T

Tetraethyl lead, 17, 127–128
Thermal reactor, 55
Three-way converter, 65–69, 97
Throttle body fuel injection, 183
Transmission vacuum switch, 78–79

V

Vacuum advance, 107–116
Vacuum amplifier, 74, 86–87
Vacuum delay valve, 113–116
Vacuum motor, 42–43, 46